# Geometrical Quilts

*Pat Storey*

*Tarquin Publications*

# Before You Begin



# Section 1

▲ = beginner with some experience of techniques ▲▲ = intermediate level ▲▲▲ = advanced (or intermediate wanting to try something challenging)



* *Variation shown from a Computer Aided Design program.*

** *Variation shown in totally different colour schemes – actual quilts.*

** *Variation shown in completely different colour schemes – actual quilts.*

# Before You Begin

Although these quilts are presented as a series of small quilts – wallhangings, there are many ways in which they can be used – almost like blocks – to make larger hangings and bed quilts. Instructions for making some larger quilts are given in Section 3 of the book. Also, variations to the basic design are often possible. Some of these are given within the individual chapters.

**A quarter inch seam is used throughout.**

## Techniques

As this book is primarily 'how to make', rather than 'how to do', basic knowledge of the many techniques used for the quilts is presumed. Full construction details are set out and any other explanations felt to be helpful will be given alongside them. The Appendix will include names of books and websites of interest to anyone needing additional information.

The quilts are graded, according to difficulty, in the contents list. For the purposes of these quilts, 'Beginner' is someone with little experience but who has mastered basic skills of accurate cutting and sewing. To have learned the technique of Foundation Paper Piecing (FPP) would be especially useful for the designs in this book, though there are several patterns which do not use it.

## Fabric

I have used 100% cotton throughout. The quantity of fabric required for the quilt tops depends on so many variables – how many colours are used, how they are distributed (for example, you might wish to duplicate a colour from one quilt to another), whether the quilts are enlarged or reduced in size and so on – that it is not possible to give precise yardage for each quilt. As I guide, I can say that I did not need more than a quarter of a metre (yard) of any individual fabric, for the designs themselves. If you wish to make the backgrounds, of any centre piece, larger than 18″ square, you will need a piece a little larger than a fat quarter for these. The section of the background which is underneath appliqué may be cut away for re-use, so that it isn't 'wasted'. I don't generally specify a 'fat quarter', as long quarters are often useful for borders and binding. I often found it possible to use pieces from my (not vast) stash in the quilts. I have included a guide to fabric amounts for the large quilts in Section 3.

Pieces of fabric for backing the quilts will also be needed. The largest of my examples, which is based on a quilt centre of 20″ square, requires a piece of fabric no less than 26″ square. Batting sizes will match those of the backing.

## Colours

The examples I have made, shown throughout the book, show only one example of the possible colour schemes/numbers of different colours you could use. A few variations, made by friends, are included. I hope these will give you an idea of the difference that more colour can make.

## Thread

I am not dictatorial about thread, though I now use a fine, long staple cotton thread for piecing (having used up my polyester thread in my early years of patchwork and quilting). For quilting, I often use a variegated thread, sometimes one of the lovely rayon threads available. For a bed quilt, which will have harder use than a wallhanging, I would probably use a cotton quilting thread; or, indeed, one of the interesting fancy threads about nowadays – perhaps quilting from underneath. I have used a modern 'invisible' thread (polyester rather than nylon), when I want the design to retain more impact than the quilting.

## Batting

I used Hobbs Fusible Batting for all these quilts. I use this more than any other for wall hangings. It has a fairly low loft and hangs beautifully flat. It is easy to apply using a steam iron (I hold the iron slightly above the surface of the quilt rather than applying pressure). I do steam top and back, and also fasten the outside edges with basting pins or tacks. This is a 'belt and braces' approach.

## Quilting

My quilting is all machine quilting. I have included suggested quilting designs for each of the small quilts, if you wish to use them. I do not mark my quilts (no law, just preference); so my designs are based on free-motion quilting, stitch in the ditch, patterns where the side of the presser foot can be used as a guide, freezer paper templates (quilted around), or drawing the quilting design on fine paper (such as Golden Threads), pinned to the quilt and sewn through.

If you are like many who do not feel they have mastered the art of free-motion quilting, I must admit to being in the same category myself. However, as a word of encouragement, let me say that I was a lot better at the end of the series than I was at the beginning!

## Mitred Corner

The attached diagrams show the main stages in sewing a mitred corner – most often used with borders, wide or narrow.

First stage is to sew one border piece to the 'receiving' piece, starting and finishing the seam a quarter of an inch before the edges of the 'receiving' piece. Press this border away from the 'receiving' piece.

Second stage is to sew the second border piece to the receiving piece, starting and finishing a quarter of an inch before the edges of the 'receiving' piece. Do not fold this piece back yet.

You will see that there is a point where the outer edge of the second border piece, lying (wrong side up) over the first border, meets the outer edge of the first border piece, perpendicularly.

Draw a pencil line between this point and the end of the second seam.

Pick up the pieces and align the edges of the border pieces, tucking the 'receiving' piece out of the way between them. Pin, so that the edges remain firmly together. Sew along your drawn line. Unfold, to check that all is correct, then trim the mitre seam and press it open. Don't cut before checking! Press the remaining border piece away from the centre.

(The third and fourth border pieces are sewn on in the same way).

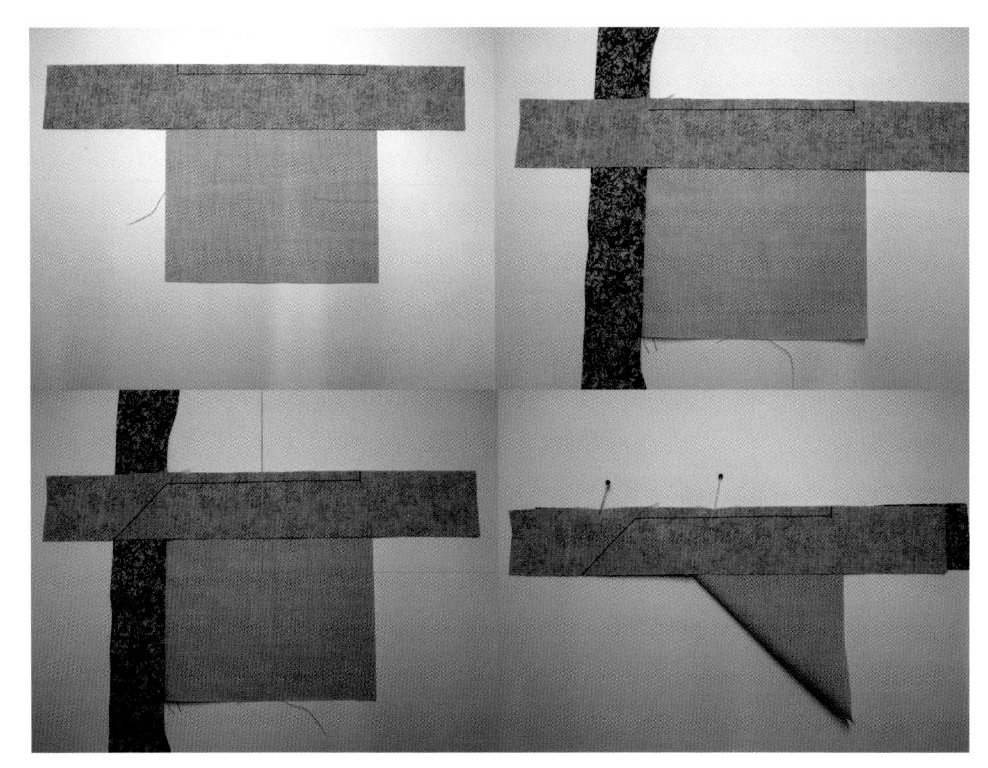

*Mitred corners*

## Binding

I use the single, continuous strip method of binding with mitred corners. There is no need to use bias strip for these square/rectangular quilts, though some people prefer bias whether the quilt is straight or curved. I find it easier to achieve neat corners with the mitred method than with the 'sides then top', squared corners method. But, if you prefer this way, do stick with it. I have used single thickness, quarter inch binding for these small quilts; but, for a bed quilt, I would suggest using the double thickness, for strength.

## General Tips for Foundation Paper Piecing (FPP)

I say 'paper', as I prefer this medium. Alternatively, you could use a fine muslin or interfacing, if you like working with this, and the foundation would remain within the quilt.

- Start and finish each seam one stitch beyond the drawn line (in case removing the paper loosens the first stitch).
- Use a very fine, fibre-tipped pen for drawing to avoid the potential problem of having graphite dust, from a soft pencil, being transferred to fingers and then to fabric and/or thread. Printing from the computer will also prevent this.
- If you draw/trace your patterns, multiple copies can be made by using your machine, unthreaded, to perforate a stack of paper – perhaps six or eight sheets – with the master copy on top (the pack would be temporarily stapled together).
- Make an extra, numbered copy of your pattern and cut it up carefully, to use the pieces for templates. Place these pieces, right side up, on the wrong side of the fabrics and cut round them, leaving an allowance of about half an inch. This does not have to be accurately cut, but it ensures that pieces are always large enough and that angled pieces, in particular, can always be placed in the right orientation. It also reduces waste.
- Scoring every seam, with the back of a stitch ripper and a ruler, makes removing the foundation paper perfectly easy.
- Using a smaller stitch than usual is always recommended; but don't make it too small, as this makes unpicking very much more difficult – should you ever have to do any!
- You will notice that some of the patterns given for FPP are shown with a quarter inch seam allowance drawn in; but others are shown without this. The reason for the difference is that, where two or more FPP sections are to be joined together, it is much easier to do this without having paper within the seam allowance. For these, when you have finished the piecing, trim down to the required size by setting the quarter inch line on your ruler to the exact edge of the paper pattern. The fabric will thus be cut an accurate quarter inch from the edge of the paper. There will be no paper in the way of your seam. The great advantage to this is that you are able to leave the paper on your piece until all the sections have been joined. In the pattern notes, this is described as 'Trim to size ***adding*** seam allowance'. Where the seam allowance is included in the drawn pattern piece, the trimming is described as 'Trim to size ***including*** seam allowance'. For this, you place the edge of your ruler along the edge of the pattern in the usual way.

## Grain Lines

Borders and binding should all be cut on the straight grain of the fabric; and anywhere else, where straight strips are used, e.g. the Tessellation quilt. Where a chosen fabric has an obvious, directional pattern, the direction should be decided in advance and all the pieces using that fabric, within the quilt, should be cut according to that decision. Where the fabric does not have a directional pattern, and FPP is being used as the technique for piecing, cutting in a particular grain direction is not so important, as the paper stabilises the fabric, so that potential problems with bias do not arise. Wherever large single pieces of fabric are used, e.g. the plain triangles in Sierpinski's Triangle, the appearance of the quilt will be enhanced by cutting them in the same orientation (with a tone-on-tone sort of fabric, the triangles may be cut side by side so that the spaces between are also used – the fact that the fabric of one is upside down to the fabric of another will not matter with such a fabric). Using acute-angled triangles, as many of these designs do, two sides will never be on the straight grain. When I am cutting triangles, I usually try to ensure that the baseline is cut on the straight grain.

## Borders and Their Quilting Patterns

The suggestion for quilting many of the borders of the small quilts is for free-motion quilting of various kinds. However, the suggestions for the remaining borders involve patterns which have been drawn to certain measurements. Where the centre piece of the quilt is appliquéd to a background, the final trimming of the background, once the centre piece has been added, can be such that the border quilting pattern can be fitted in exactly.

Where the centre part is all pieced, however, some small adjustments to the size as given in the pattern might be needed. The way to do this is, as follows:

Measure your border. The quilting instructions will give the number of pattern elements you might expect to fit into your border. Divide your measurement by this number. The answer will give you the size you need for each repeat. You can then re-draw the pattern making whatever adjustment is needed. The pattern or the quilting instructions (found at the end of the construction section) will suggest where it will be best to make the adjustment.

## Cornerstones

Patterns for the cornerstones used in all the small quilts have been grouped together within the Pattern Pack. They were all designed to have some connection with the design of the quilt to which they 'belong'. However, there is no reason why they should not be interchanged if you wished to do so. This is why they have been presented as a group in their own right. Instructions for making the Cornerstones for the Baravelle Spiral quilt are given with the other instructions for that quilt, as they are not made with Foundation Paper Piecing.

## Pressing

I have found that using a dry iron for pressing during the process of making any block, or quilt in this case, sets the pieces sufficiently well to give accurate results. When the piece is finished, I like, then, to use a steam iron to get really sharp seam lines. Traditionally students are taught to press seams first of all to the darker fabric; then, because that is not always convenient, to press to the side. But I am now finding more and more instances where pressing seams open gives a more pleasing result (without the otherwise inevitable 'lump') and also where it makes the seams easier to manage. For instance, quilting by the stitch in the ditch method is much easier with the seams pressed open. Keep an open mind, experiment with sewing and pressing a couple of pieces of scrap fabric together and see what result you prefer.

Another effect you can achieve simply by directional pressing is to make one piece stand out from the surrounding pieces. If all the seams surrounding that piece (perhaps a 'fussy-cut' picture?) are pressed underneath it, a definite raising of the piece results – a very small amount, of course, but noticeable.

## Templates

Several of the patterns in this book use templates. Some of these are of freezer paper, where the template is pressed/stuck to the wrong side of the fabric and the seams are sewn along the edges of the freezer paper. Other instances are the more traditional use of card or template plastic. The easiest way I have found of making templates from card is to trace the pattern with tracing paper and then spray fix the traced pattern to the card ***before*** cutting it out. I have used spray adhesive marketed for photographic mount purposes, to achieve a good, even coverage. You could also use carbon paper to transfer the drawing from tracing paper to the card, if you prefer (and if you can find some!). Always remember that the sewing line is the inside of any drawn line.

## Corners

In the course of making these quilts, you will sometimes find it necessary to mark the point where seam lines will intersect at the corners. The diagrams below can be made into templates for marking corners with four of the most common angles. Using this same principle, you can make a template for any angle you might require at any time.

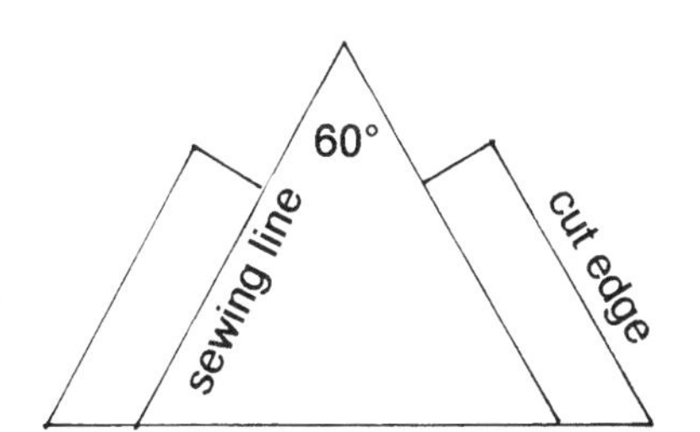


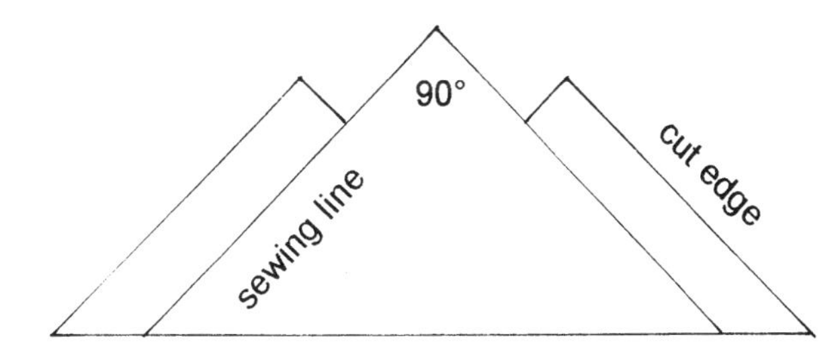


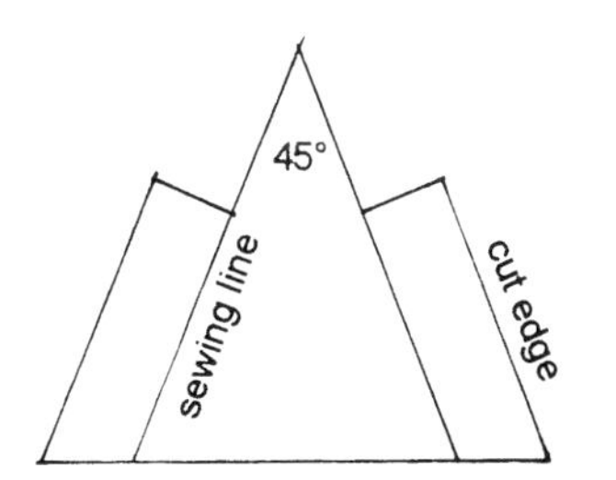


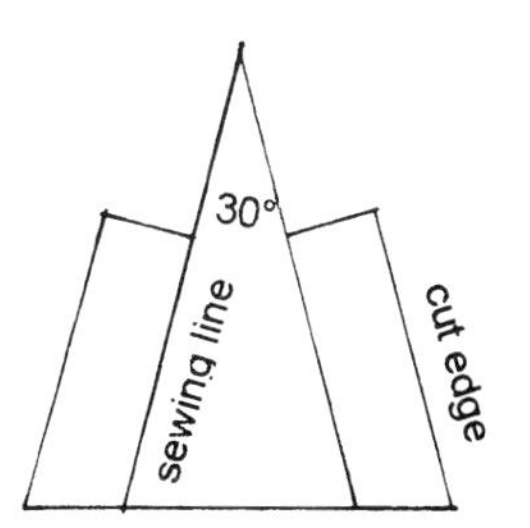

# Chapter 1 – SPIRAL TRIANGLES

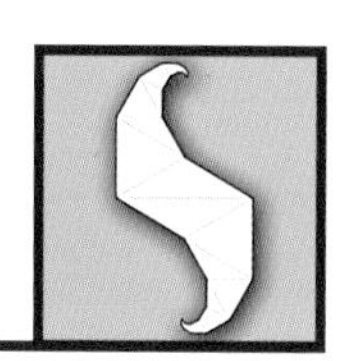

*The illusion of a spiral is created when similar, equilateral triangles are repeatedly rotated, within one another, at a regular interval. There are no curved lines, but the appearance of curves is convincing. The smaller the interval of separation, the better the curves appear.*

## Fabric Selection

- Three contrasting colours – one for each of the three sectors of each triangle: Fabrics 1A, 1B and 1C. Also, one tiny piece for the centre triangle. I used the background colour, but it can be anything you choose.
- Fabric 2. Background.

**Plus:** *if you are making the complete small quilt:*

- Fabric 3. For the narrow border and binding. (Cut 1″ and 1¼″ wide strips respectively.)
- Fabric 4. For the wider border. This can be the same as Fabric 2, if you like the idea of the background extending beyond the narrow border. (Cut 2½″ wide strips.)
- Scraps of fabrics used elsewhere in the quilt, for the pieced cornerstones.
- Backing fabric. A piece slightly larger all round than the finished quilt will be.

## Assembly Instructions

1. Cut 1½″ strips of fabric for the Foundation Paper Piecing. These will be ample for all pieces. There will be some waste cut away, but cutting actual wedge shapes of many different lengths – to avoid the waste – would be time-consuming and restrictive. I do not recommend this cutting plan.
2. Make three, pieced triangles from Pattern A (clockwise rotation) – *Fig.1,i (PP)* and three from Pattern B (anti-clockwise rotation), *Fig.1,ii (PP)*. Trim to exact size including the quarter inch seam allowances and remove papers. Press.
3. Cut two setting triangles from the pattern as shown – *Fig.1,iii (PP)*; plus, because the triangle is not symmetrical, cut two from the pattern reversed.

**4.** Arrange the pieced triangles, alternating the direction of rotation, and so that like-coloured sections are adjacent, forming the required hexagon.

**5.** Sew the pieced triangles together in sets of three, alternating the direction of rotation; then put the two halves together, as diagram – *Fig.1,iv*, carefully matching the points where the colours change. Press.

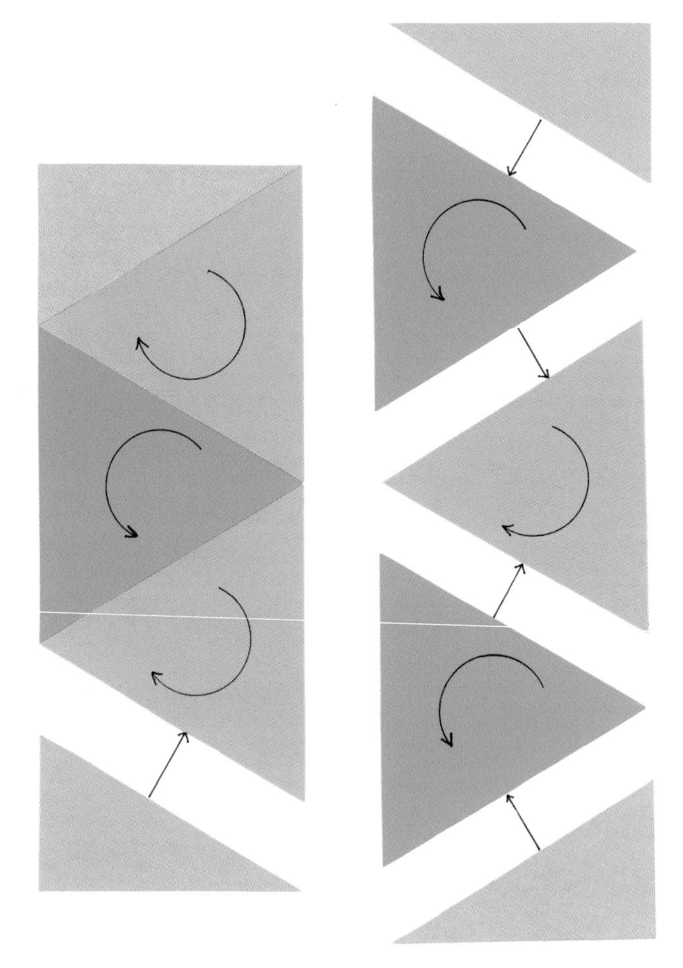

*Fig.1,iv*

**6.** Sew the setting triangles to the hexagon so as to create the rectangular shape.

- As the setting triangles are not symmetrical, you will need to place the pairs of triangles, as pattern and reverse, on diagonally opposite corners.
- If you plan to use this centre piece as a block in a large quilt, such as those described in Section 3, the work for this design is now complete.
- If, however, you are making a wallhanging or similar small quilt, the finishing instructions, which are common to all of the designs, are given at the end of this section.
- Possible quilting designs are given below – Figs.1,v–viii.
- See also full quilting details at the end of this section.

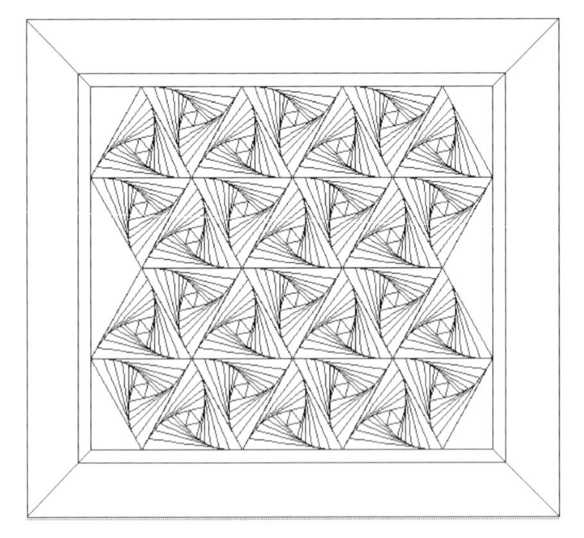

*Created by Tutu, using EQ5*

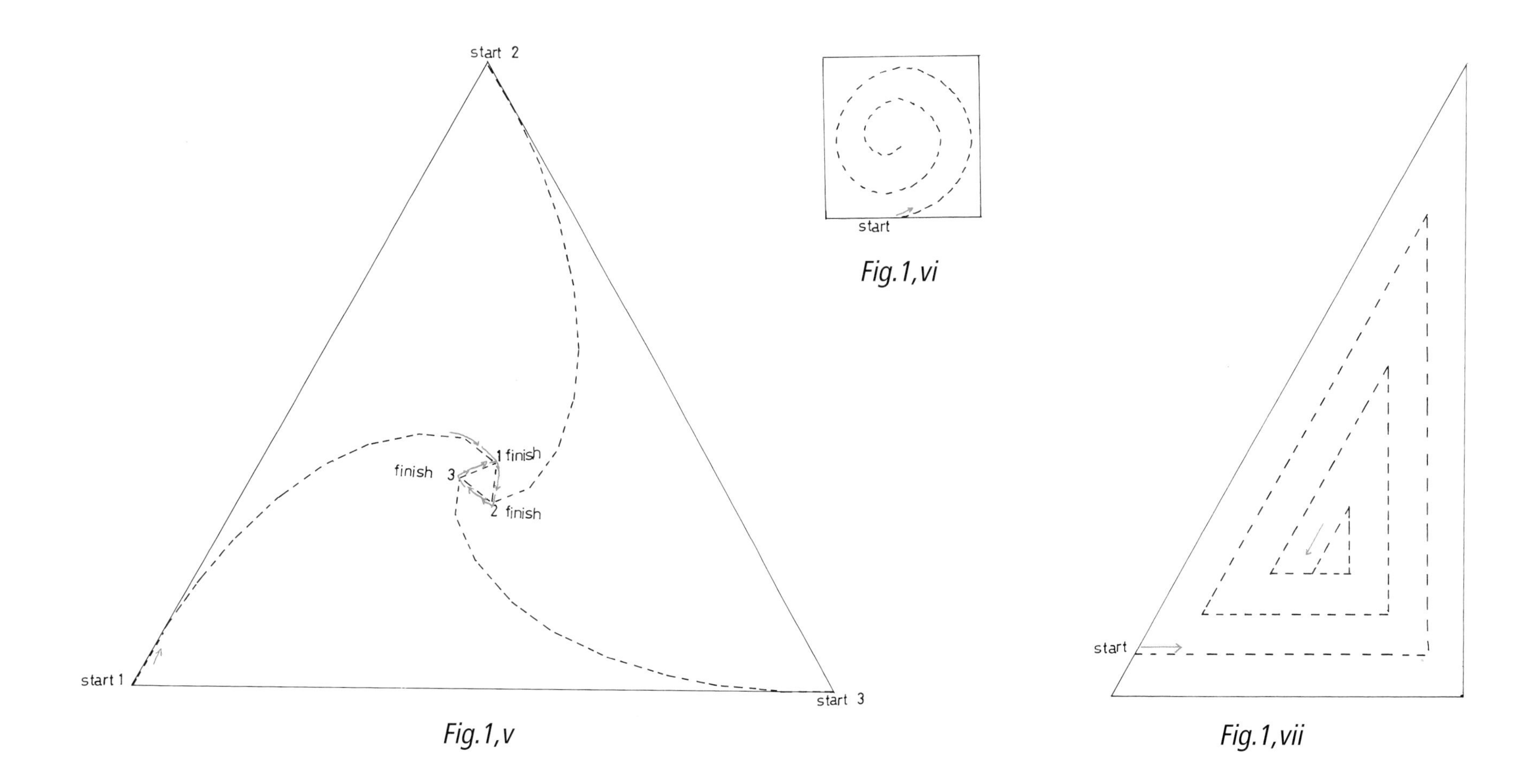


*Fig.1,v*

*Fig.1,vi*

*Fig.1,vii*

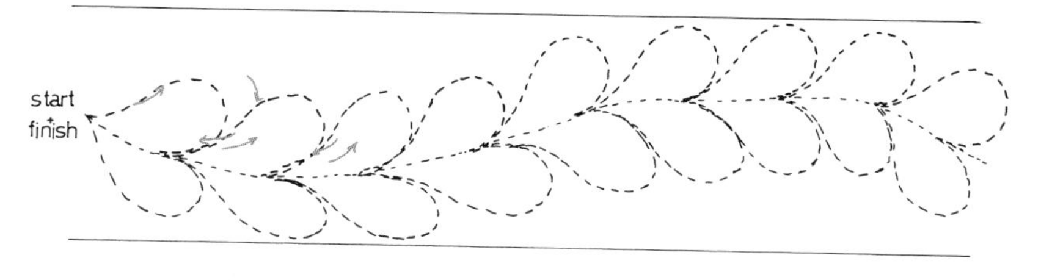


*Fig.1,viii*

# Chapter 2 – LOGARITHMIC SPIRAL

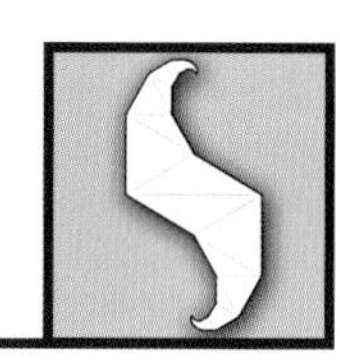

*The Logarithmic Spiral (also known as the Equiangular Spiral) may be explained thus: a spiral which, in its growth from the centre, widens and lengthens in unvarying proportions, retaining its shape at every stage.*

## Fabric Selection

You will need:

- Fabrics for the spiral itself. You will need a minimum of two, contrasting fabrics. You can use more, and decide on your own colour placement. In the alternative version shown, a different fabric has been used for every piece. It is important to remember that, in order for the spiral to be clearly defined, fabric pieces need to be distinguishable alternately ***within*** the wedges and from the pieces they will 'meet' on either side. Fabric 1A, 1B, and so on.

  A completely different effect, from my example quilt, is the one shown below. This is what can happen when 70 different fabrics are used!

- Fabric 2. For the background piece. This needs to be about 1" larger, all round, than the size you want it to be when finished.

**Plus:** *if you are making a complete, small quilt:*

- Fabric 3. For the narrow border and binding. (Cut 1" and 1¼" wide strips respectively)
- Fabric 4. For the wide border. This can be the same as Fabric 2, if you like the look of the background extending beyond the narrow border. (Cut 2½" wide strips)
- Small pieces of fabrics used elsewhere in the quilt, for the pieced cornerstones.
- Backing fabric. A piece slightly larger all round than the finished quilt will be.

## Assembly Instructions

1. Trace the spiral, from the pattern – *Fig.2,i(PP)*, onto paper suitable for foundation paper piecing. You will use this tracing to paper piece the wedges.
2. Cut apart the spiral's wedges very carefully. It is not necessary to leave seam allowances. It is a good idea to number the wedges at this stage.
3. Decide on the placement of your chosen colours.

4. Piece the wedges, using the paper foundations. Press seams. At the pointed end, carefully draw round the paper onto the reverse of the fabric – use a ruler, to help hold the paper down close to the fabric. This will help to get the centre of the piece accurately positioned – important when there are so many seams joining at the one place.

5. Trim the wedges adding a quarter inch seam allowance. You may remove the papers at this stage if you wish.

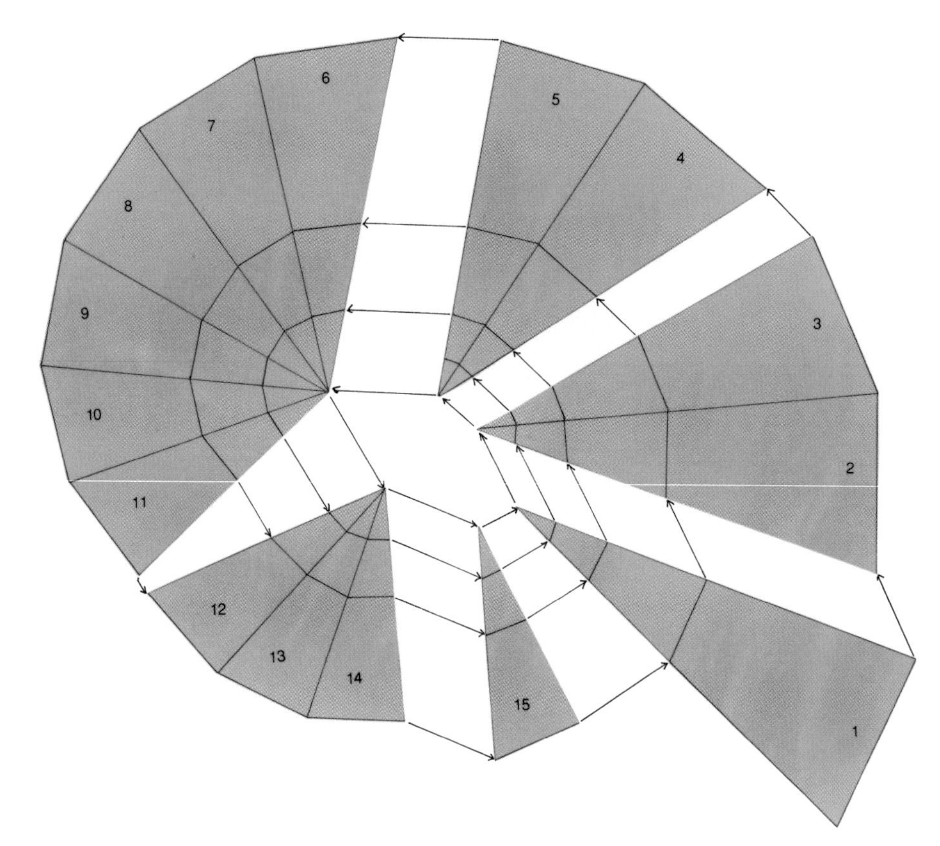


*Fig.2,ii*

6. Join the wedges, in adjacent pairs first, matching all 'horizontal' seams – *Fig.2,ii.* When pinning the pieces together, if you have retained the paper, pin only into the fabric beyond the paper to avoid distortion.

7. When joining the pairs of wedges, and with all subsequent joining, use the drawn lines at the centre end of the wedges in order to match the pieces exactly. Start all these joining seams at the centre and sew outwards.

8. Join the pairs to make larger groups then, finally, eight and seven. Finally, join the eight wedge and the seven wedge sections. Press. Remove papers now, if you did not do so earlier.

9. Along the outer edge, press the quarter inch allowance under.

10. Appliqué the spiral to your chosen background fabric. It will need to be pinned, and/or held with dabs of fabric glue, to prevent movement during the appliqué process. It may be hand or machine appliquéd. If you decide to machine appliqué, I suggest using invisible thread and a tiny zig zag or narrow blind hem stitch. Press. Trim and square this centre piece to the required size.

    - If you plan to use the centre piece as a block in a large quilt, such as those described in Section 3, the work for this design is now complete.
    - If, however, you are making a wallhanging, or similar small quilt, the finishing instructions, which are common to all the designs, are given at the end of this section.
    - Possible quilting designs are given below – *Figs.2,iii and iv.*
    - See also full quilting details at the end of this Section.

*Fig.2,iii*

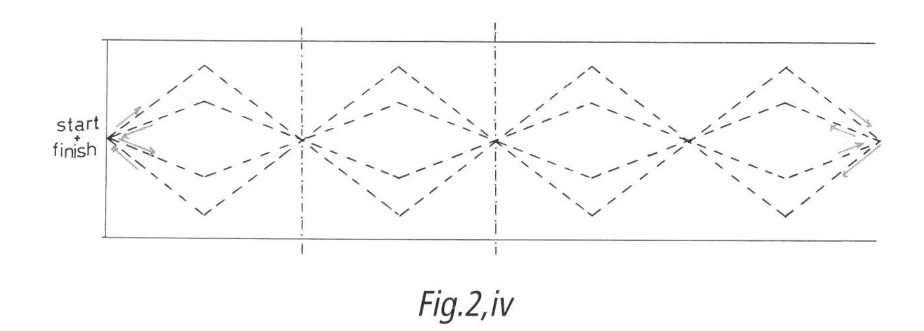


*Fig.2,iv*

*Helen's Alternative Version*

# Chapter 3 – BARAVELLE SPIRAL

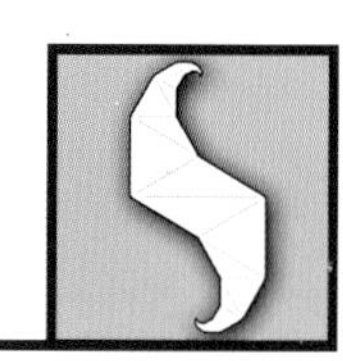

*The Baravelle Spiral is a fractal and a 'straight line' spiral. The curved effect derives from the way the structure is drafted. The Spiral may be drawn from any regular polygon. Repeated joining of the mid-points of the sides of the polygon, to create similar polygons within the original, achieves the required base. The spiral becomes apparent when a sequence of triangles is coloured or shaded. The shading starts with one of the outermost triangles, then to the adjacent triangle in the next layer (clockwise or anti-clockwise) and so on, to infinity – or until it is not physically possible to continue. Pentagon, hexagon and heptagon perhaps produce the clearest spirals.*

## Fabric Selection

- For the central hexagon, I used five different fabrics, repeating the darkest, to highlight the spiral effect. The remaining four spirals are made with four lighter fabrics, in toning colours. Fabrics 1A, 1B, 1C, 1D and 1E.

- Fabric 2. For the background. Choose a fabric which offers good contrast to the spiral design itself.

**Plus:** *if you are making a complete, small quilt:*

- Fabric 3. Narrow border and binding. (Cut 1" and 1¼" wide strips, respectively.)

- Fabric 4. For the wider border. This can be the same fabric as Fabric 2, if you like the look of the background extending beyond the narrow border. (Cut 2½" wide strips.)

- Small pieces of fabrics used elsewhere in the quilt, for the pieced cornerstones.

- Backing fabric. A piece slightly larger all round than the finished quilt will be.

## Assembly Instructions

**1.** Copy or trace the pattern – *Fig.3,i (PP)* onto a large sheet of lightweight paper, suitable for Paper Piecing. Colour or number the drawing – *Fig.3,ii* so that it is clear where all your fabrics are going to go. Numbers on the pattern show the order in which it should be pieced; letters refer to the way the various colours form the spirals. This is a single piece of Foundation Paper Piecing; and, as it is a fairly large piece, might otherwise have a tendency to confuse – due to the combined nature of the shape and the reverse technique of

Foundation paper piecing. I used flipchart paper. Smaller sheets could be joined, but this will add to bulk.

**2.** I strongly recommend pre-cutting all fabric pieces to shape and approximate size (always about half an inch over size on all sides) before you begin to piece. The pieces should be placed together according to colour and size (smallest on top). This will enable you to pick the correct piece from the top of each pile as you go round.

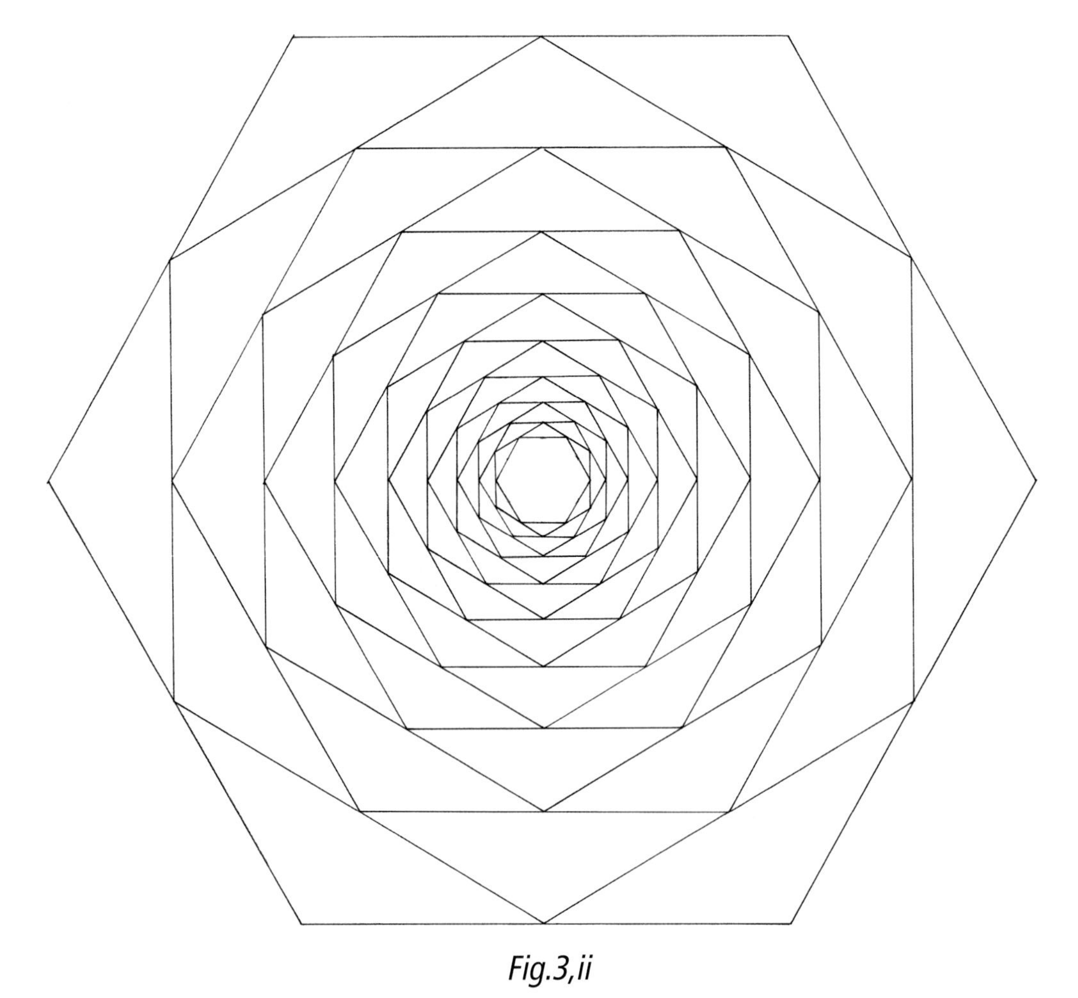

*Fig.3,ii*

**3.** Pin the centre piece of fabric in place. Begin to add the pieces, in strict numerical order. Once the first few are in place, and the pins are removed, the rest of the process is straightforward. You need to take extra care at the beginning, while the pieces are small and the remaining paper is large. Finger press each seam as you complete it. Press with an iron after every couple of revolutions. When this hexagonal part is complete, press and trim to the exact size adding quarter inch seam allowances.

**4.** Cut four background pieces, using templates made from the pattern given – *Fig.3,iii (PP)*. As these setting triangles are not symmetrical, two should be made from the pattern as drawn; the template must be reversed for the second pair.

**5.** Sew these background pieces to the hexagon, to create the rectangular shape required. Pairs should be sewn to diagonally opposite corners – *Fig.3,iv*. Trim and square this centre piece to the required size.

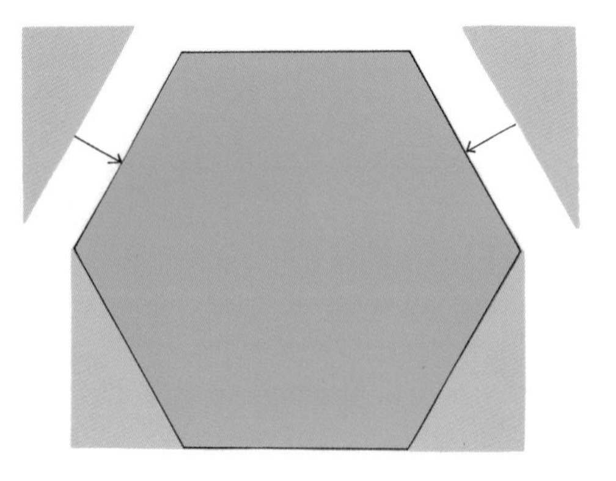

*Fig.3,iv*

- If you plan to use the centre piece as a block in a large quilt, such as those described in Section 3, the work for this design is now complete.
- If, however, you are making a wallhanging, or similar small quilt, the finishing instructions, which are common to all the designs, are given at the end of this section.
- Possible quilting designs are given below – *Figs.3,v-vii.*
- See also full quilting details at the end of this section.

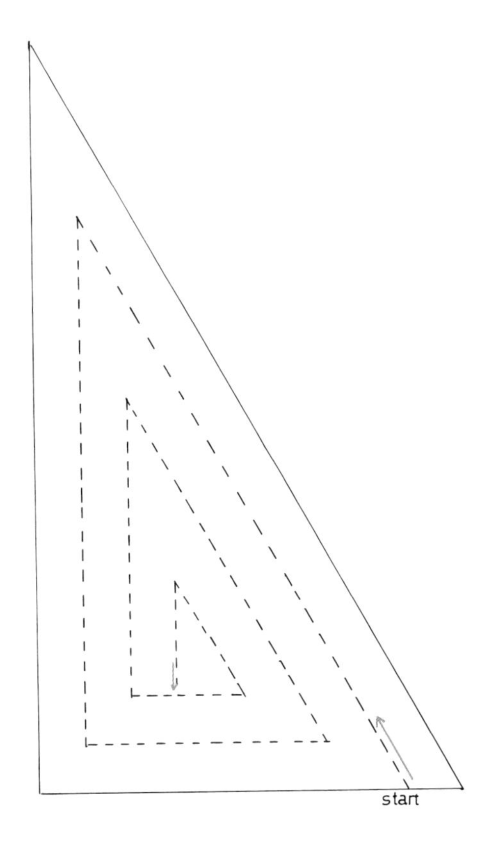


*Fig.3,v*

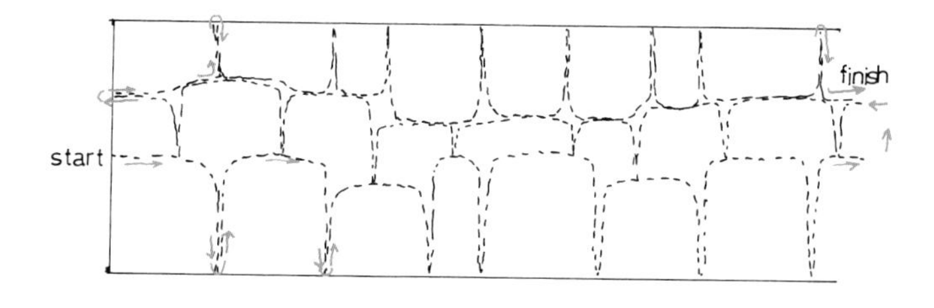


*Fig.3,vi*

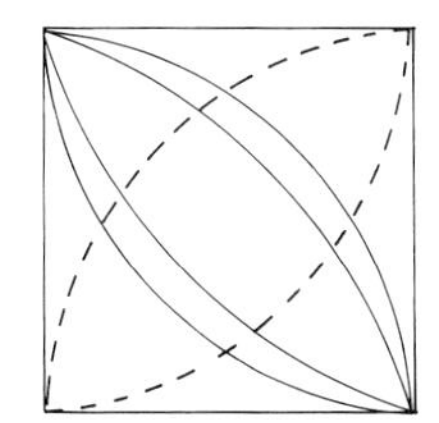

*Fig.3,vii*

- The Cornerstones for this small quilt are not foundation paper pieced. They are made as follows:
- Three 2½" squares of fabric are required for each corner.
- Fold two of the squares in half, on the diagonal and tack/baste them to two diagonally opposite corners of the third, background square.
- Roll back the bias edge of both the top two squares. Pin to hold, and stitch down the rolled back edge. I always hand stitch this technique. When both are done, you have the elliptical shape.

# Chapter 4 – ARCHIMEDEAN SPIRAL

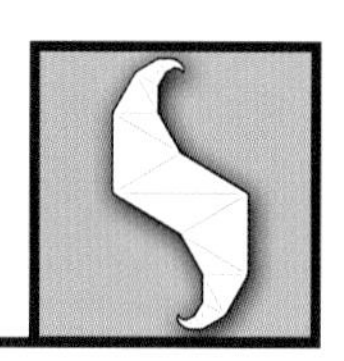

*An Archimedean Spiral is one in which the breadth of the whorls remains the same ("constant separation distance") however large the spiral becomes.*

## Fabric Selection

- Five fabrics, of good contrast with one another, for the centre plus the four concentric bands which make up this spiral pattern. Fabric 1A, 1B, 1C, 1D and 1E. Repetition of a fabric would be fine in this design, provided the spiral bands remained distinguishable.
- Fabric 2. Background fabric which contrasts well with the outer band. This needs to be about 1" larger, all round, than the size you want it to be when finished.

**Plus:** *if you are making the complete small quilt:*

- Fabric 3. For the narrow border and binding. (Cut 1" and 1¼" wide strips, respectively.)
- Fabric 4. For the wider border. This could be the same as Fabric 2, if you like the idea of the background extending beyond the narrow border. (Cut 2½" wide strips.)
- Small pieces of fabrics used elsewhere in the quilt, for the pieced cornerstones.
- Backing fabric. A piece slightly larger all round than the finished quilt will be.

## Assembly Instructions

The centre piece of this quilt is made in quadrants.

1. Trace the individual circular bands from the pieces in the Pattern Section, to make templates – *Figs.4,i-iv (PP)*. Number your quadrants, and the component bands of each quadrant. Keep all the pieces of each quadrant together.
2. Decide on your colour placement, and draw round the templates on the wrong side of each of the fabrics in your selection. Cut these out, allowing approximately a quarter of an inch seam allowance.

3. Taking one quadrant at a time, and starting with the central piece, sew the bands together. Clipping the curves and frequent pinning – matching the positioning marks plus a few pins in between these – will facilitate the sewing of these curved pieces. You might find that the very smallest centre pieces are easier to sew together by hand. Press each seam as it is sewn. Avoid any lateral movement during the pressing, to prevent any distortion of the pieces.

4. Sew quadrants A and B together, matching seams carefully. Press. Then put quadrants C and D together. Press each seam.

5. Sew the pairs of quadrants together, matching seams carefully – *Fig.4,v*. You will discover that there is a half-inch square hole in the centre. Turn the raw edges under. Cut a 1" square of the innermost fabric and place it under the hole. Using a matching thread, and a small appliqué stitch, attach the spiral to the square (reverse appliqué). Press.

6. Cut the background fabric to your desired size plus an inch or so all round.

7. Appliqué the spiral to the background, matching thread to the colour of the outermost band. Press.

8. Square and trim this centre piece.

   - If you plan to use this centre piece as a block in a large quilt, such as those described in Section 3, your work is now complete.

   - If, however, you are making a wallhanging or similar small quilt, finishing instructions, which are common to all designs, are given at the end of this section.

- Possible quilting designs are given below – *Figs.4,vi-viii.*

- See also full quilting details at the end of this section.

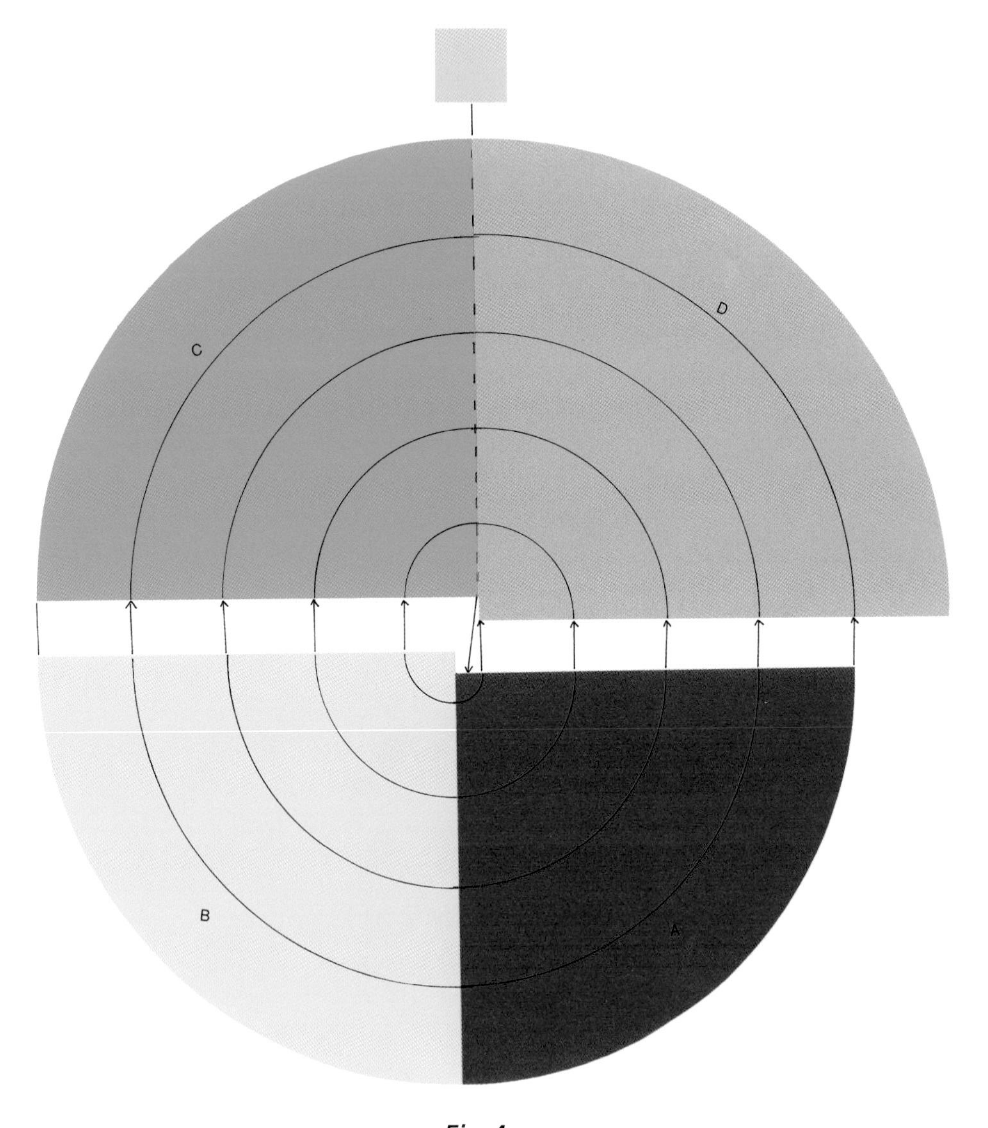


*Fig.4,v*

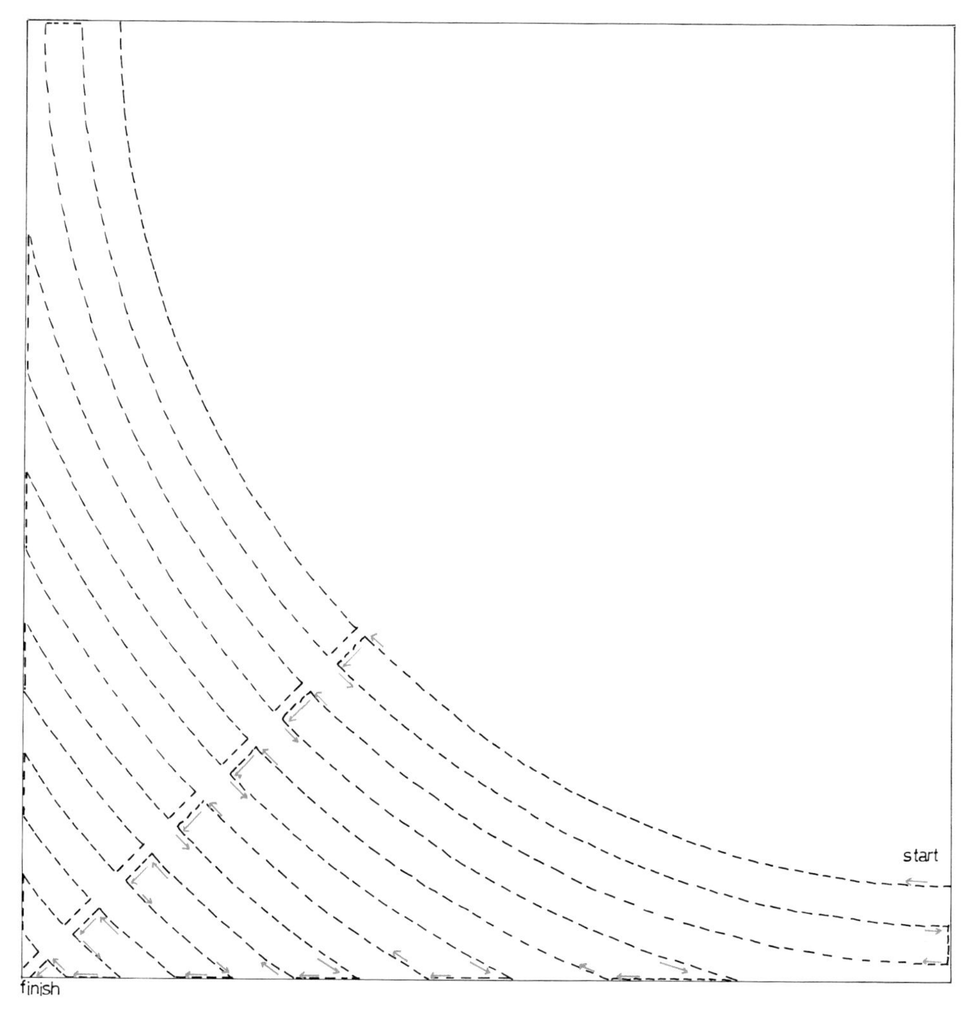


*Fig.4,vi*

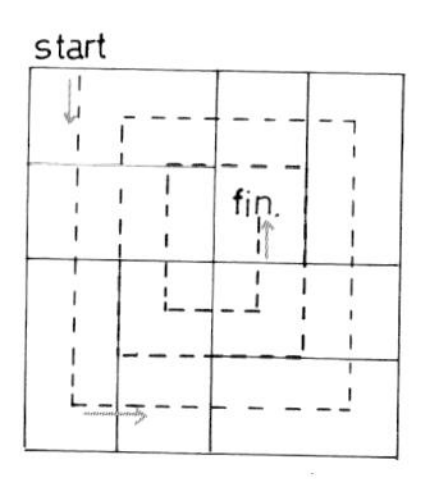


*Fig.4,vii*

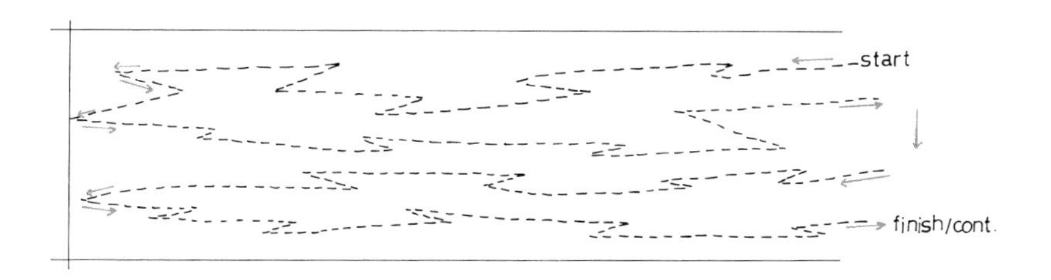


*Fig.4,viii*

# Chapter 5 – SIERPINSKI TRIANGLE

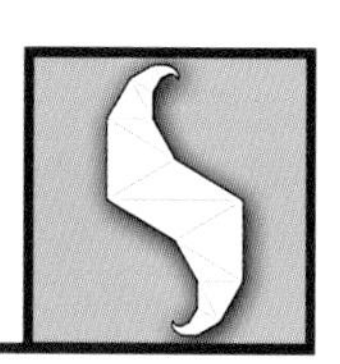

*A fractal discovered by Waclaw Sierpinski (also known as Sierpinski's Gasket). It is shown here using an equilateral triangle. The principle is that a triangle may be repeatedly shrunk to half its size, re-drawn within itself, to infinity. In the illustrated quilt, the main triangle shows four iterations of the division; the background, three.*

## Fabric Selection

- Fabrics 1A and 1B. Two contrasting colours for the main triangle.
- Fabrics 2A and 2B. Again, two contrasting colours, different from the first two. I used less bold colours for the background than for the main triangle.
- Fabric 3. For the narrow, outlining strip to the main triangle. (Cut 1" wide strips) *(May be omitted.)*

**Plus:** *if you are making the complete, small quilt:*

- Fabric 4. For the narrow border and binding. This can be the same as Fabric 3 (if used), or closely related. (Cut 1" and 1¼" wide strips, respectively.)
- Fabric 5. For the wider border. (Cut 2½" wide strips.)
- Small pieces of fabrics used elsewhere in the quilt, for the pieced cornerstones.
- Backing fabric. A piece slightly larger all round than the finished quilt will be.

## Assembly Instructions

1. Trace or print 27 copies of the small Foundation Paper Pieced triangle pattern in the Pattern Section – *Fig.5,i (PP)*
2. Trace or print 8 copies of the larger Foundation Paper Pieced triangle pattern in the Pattern Section – *Fig.5,ii (PP)*
3. From the remaining triangle patterns in the Pattern Section, make templates for the main triangle, as follows:

   From Fabric 1A, cut 1 of the largest size – *Fig.5,iii (PP)*, 3 of the medium size – *Fig.5,iv (PP)* and 9 of the small size – *Fig.5,v (PP)*.
4. From Fabric 2A, for the background, using the same templates as for the main triangle, cut 2 of the medium size – *Fig.5,iv (PP)*.

   Cut 2 large half-triangles – *Fig.5,vi (PP)*, (reverse the template for the second one)

Cut 2 medium half-triangles – *Fig.5,vii (PP)*, (reverse the template for the second one)

5. Make up the 27 small Foundation Paper Pieced triangles in Fabrics 1A and 1B. I suggest pre-cutting all the component triangles: 27 of Fabric 1A and 81 of Fabric 1B. Press all completed triangles. Trim to size adding the seam allowance. Do not remove the papers at this stage, because there are so many bias edges.

6. Build up the Main triangle, as follows: – *Fig.5,viii*

*Fig.5,viii*

**i.** Sew 1 completed pieced triangle to each side of the 9 smallest, Fabric 1A, plain triangles. Press. This results in 9 medium sized triangles with a plain centre. Check for size accuracy.

**ii.** Sew 1 of these completed medium triangles to each side of the 3 medium-sized, Fabric 1A, plain triangles. Press. This results in 3 large triangles with a plain centre. Check for size accuracy.

**iii.** Sew 1 of these completed large triangles to each side of the largest plain, Fabric 1A triangle. Press. Check for size accuracy.

This makes up the actual Sierpinski Triangle. Papers can be removed at this stage.

7. Measure the side length of this Main triangle.

Cut two 1" wide strips from Fabric 3 x this measured length plus 1".

8. Sew these strips to each side of the Main triangle (not the base). Press.

*(I suggest omitting this outlining strip – i.e. Steps 7 and 8) – if you are using this design as a block in a larger quilt.)*

9. Now, on to the Background triangles. (These are actually two halves of a single, three-stage Triangle, placed upside down.)

Make 8 of the medium sized Foundation Paper Pieced patterns – *Fig.5,ii (PP)*, in Fabrics 2A and 2B.

Make 1 each of the two medium-sized, half-triangles – *Figs.5,ix and 5,x (PP)*, with the same colours and placements.

Trim all to size, adding the seam allowance. Press.

**10.** Build up the Background Triangles – *Fig.5,xi*, as follows:

- **i.** Sew 1 pieced triangle to each side of the medium-sized plain triangle. This results in a large triangle, with a plain centre. Check for size accuracy. Press.
- **ii.** Sew this triangle to the inner side of the large, plain half-triangle. Press.
- **iii.** Sew the fourth of the remaining pieced, medium-sized triangles to the inner side of the plain, medium-sized, half-triangle. Press.
- **iv.** Sew this combination piece to the base of the large, plain half-triangle. Press.
- **v.** Sew one pieced half-triangle to the base of the medium-sized plain half-triangle.

*Fig.5,xi*

Repeat as mirror image for the second background triangle.

**11.** Sew the two background triangles to the main triangle. Press.

- If you plan to use this centre piece as a block in a large quilt, such as those described in Section 3, your work is now complete.
- If, however, you are making a wallhanging or similar small quilt, the finishing instructions, which are common to all the designs, are given at the end of this section.
- Possible quilting designs are given below – *Figs.5,xii–xiv.*
- See also full quilting details at the end of this Section.

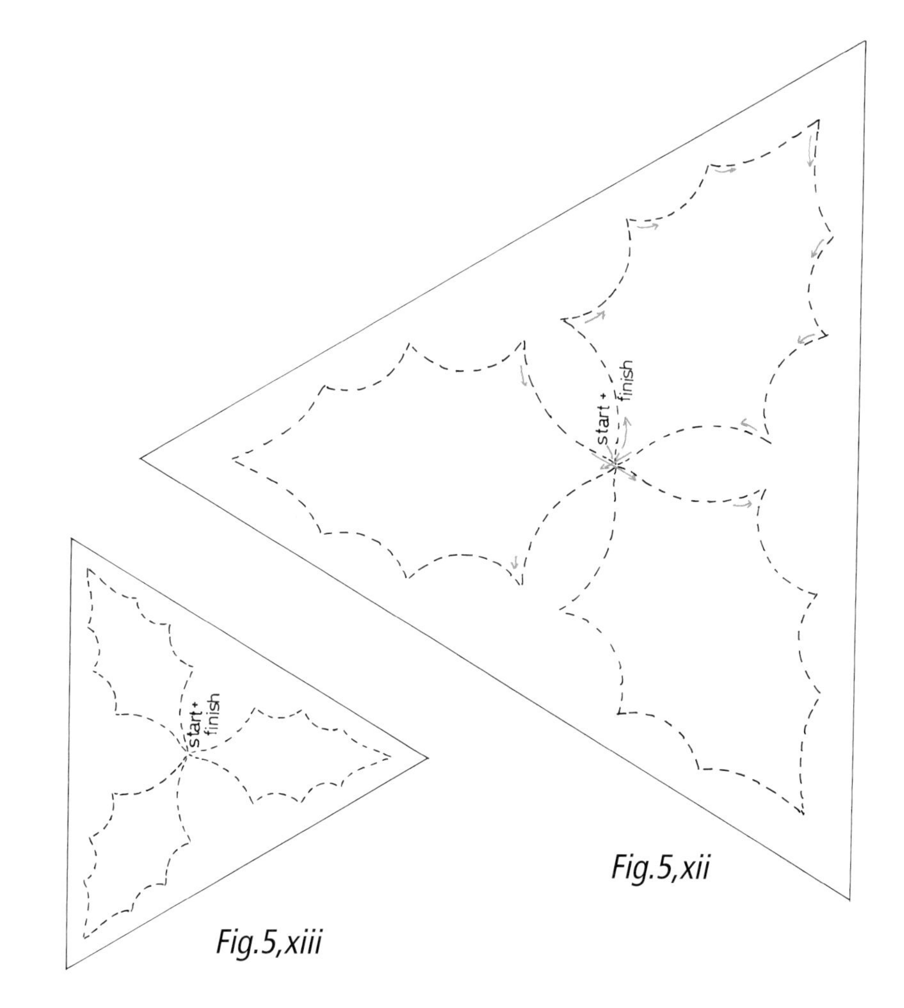


*Fig.5,xii*

*Fig.5,xiii*

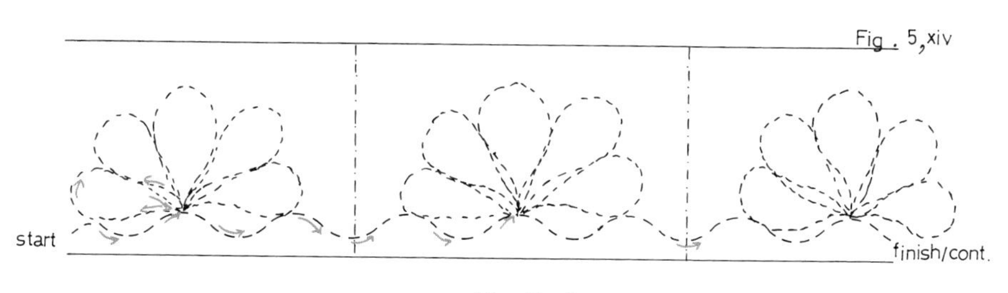


*Fig.5,xiv*

# Chapter 6 – PYTHAGOREAN TREE

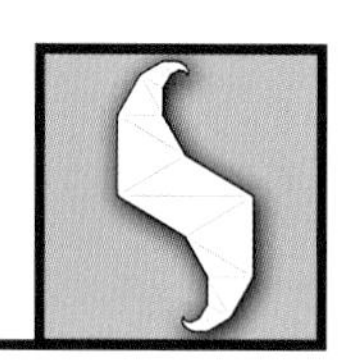

*The Pythagorean Tree is a fractal constructed from repeated sets of three squares: one base square plus two whose sides are ½ times the square root of 2 times the side of the first square. When these squares are placed so that they touch one another, in the configuration often used to demonstrate the Pythagoras Theorem, they enclose an isosceles, right-angled triangle. My example shows four Pythagorean Trees – of different sizes – set together, for the sake of the design.*

## Fabric Selection

*(I made a distinction between the squares and the triangles, in order to make the design appear as a stylised tree).*

- Fabric 1 A – ? Two distinct fabrics, or colour groups of fabric: one assigned to squares, the other to triangles. There are more squares than triangles – if that influences your decision.

  I have used five leaf fabrics for the main design – squares (plus another for the cornerstones); and three brown fabrics – triangles (plus another for the cornerstones).

  This quilt could be completely scrappy, if you wished, though I think the distinction between squares and triangles should always be obvious.

- Fabric 2, for the background piece, chosen to give good contrast to the tree. This piece needs to be cut at 18½″ square (19″ square if this piece will be used in a larger quilt, rather than a single, small quilt).

**Plus:** *if you are making a complete, small quilt:*

- Fabric 3, for the narrow border and binding (cut 1″ and 1¼″ strips).
- Fabric 4, for the wider border (cut 2½″ strip). This can be the same as Fabric 2, if you like the look of the background extending beyond the narrow border.
- Small pieces of fabric used elsewhere in the quilt, or closely related different ones, for the pieced cornerstones.
- Backing fabric. A piece slightly larger all round than the finished quilt will be.

## Assembly Instructions

1. Trace the Foundation patterns – *Fig.6,i (PP)* onto paper suitable for Paper Piecing, or onto a material which can be left in the quilt, such as a fine non-iron interfacing, or muslin etc. The latter will make the construction slightly easier, though it will be a little more difficult to do

the raw edge turning, as there will not be such a clearly felt edge. (There are pros and cons to each method.)

2. Cut out the Foundation patterns very accurately without any seam allowances. Piece, using your preferred foundation method. Press.

3. When pieces 1 and 2 are finished, they can be sewn together at the points shown on the patterns.

4. When pieces 4 and 5 are finished, they can be sewn to pieces 1 and 2, at the points indicated on the patterns.

5. Pieces 3, 4a and 5a remain separate and are set in place when the pieced elements are being appliquéd to the background.

6. Systematically fold all raw edges under the pattern paper/material, press and tack in place – *Fig.6,ii.* If you are using paper as the foundation, the paper must be removed before tacking. I found it helpful to fold, press, remove paper and then tack, working on just a few pieces at a time, so that the pressed folds remain crisp. Press the whole sections when completed.

**IF** you are making this piece as a block of a larger quilt, proceed with Step 7.

7. Centre the pieced trees on the background (cut at 19" or so, to allow for the appliqué work) - starting with the large combined piece. Pin carefully. Set the separate sections in the places indicated on the pattern and pin in place.

*Fig.6,ii*

As the edges have already been turned, this piece lends itself to one of the machine appliqué methods. My example was done by hand, but machine stitching with a narrow zig-zag or blind hem stitch, using invisible thread, would work well. Press.

**BUT IF** you are making this piece as a single, small quilt, please note:

Because I wanted to use a regular grid quilting pattern over the centre panel, I completed the top without the appliqué, according to the Finishing Instructions at the end of this Section, sandwiched it and quilted the grid. I then appliquéd the pieced design onto the quilted panel, as in Step 7 above. The borders can be quilted either before or after the appliqué.

- Possible quilting designs are given below – *Fig.6,iii.*
- See also full quilting details at the end of this section.

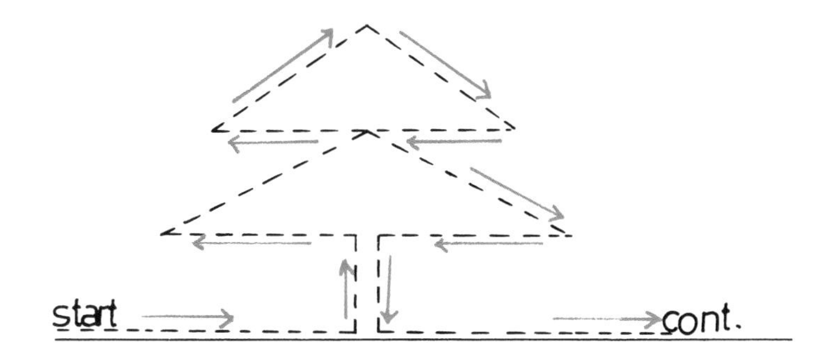


*Fig.6,iii*

# Chapter 7 – VON KOCH'S SNOWFLAKE

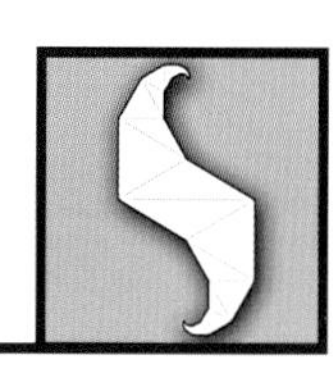

*A fractal discovered by Helge von Koch, in 1904. It is based on an equilateral triangle; constructed by notionally removing the centre third of each side and raising a further equilateral triangle on that location. At each iteration, the outline becomes more convoluted. A most interesting fact about this fractal is that, while the length of the 'snowflake' perimeter is infinite, the area enclosed by it is finite.*

## Fabric Selection

- Fabric 1 A – ? for first triangle (curved divisions)
- Fabric 2 A – ? for the second set of triangles (trapezoid divisions)

  To emphasise the fact that these three triangles are the second set, I suggest choosing completely different colour types from those of the first triangle.
- Fabric 3, for the third set of triangles (unpieced)
- Fabric 4, for the background

  Make sure that this fabric contrasts well with the third triangle set, particularly, but that it is sufficiently different from any others that are adjacent to it that no colour 'disappears' into this background colour.

**Plus:** *if you are making the small quilt or wallhanging:*

- Fabric 5, for the narrow border and binding. (Cut 1" and 1¼" strips).
- Fabric 6, for the wider border. (Cut 2½" strip).

  This fabric can be the same as Fabric 4, if you like the look of the background extending beyond the narrow border.
- Small pieces of fabrics already used in the centre part of the quilt – for the cornerstones.
- Backing fabric. A piece slightly larger all round than the finished quilt will be.

## Assembly Instructions

The quilt is assembled in sections, as diagram – *Fig.7,i.* Note 'concave' is the hollow curve (cf 'cave'); 'convex' is the opposite.

1. Trace a copy of the layout diagram – *Fig.7,i,* and colour it with your own choice of colours/fabric. After colouring, I usually stick a small square of my chosen fabric to each of the places where it will be used.
2. Section 1 – 9 triangles, sub-divided into 3 curved pieces which meet at the geometrical centre of the triangle.

*Fig.7,i*

Trace one curved piece from the pattern – *Fig.7,ii (PP)*, with which to make a template . I used tracing paper which was then stuck onto thin card and cut out carefully. Using a dry roller adhesive (or perhaps a photographic spray mount adhesive) is much less prone to distort the tracing paper than a liquid glue.

Before using your template on the fabric, check that it does indeed fit exactly over each of the three sections.

3. Following your own colour chart, and using your curved template, make the pieces for the 9 central triangles. My suggested method for doing this is to draw round the template on the wrong side of the fabric, remembering to transfer the 'matching' points. If your pencil (or other drawing implement) is in any way 'thick' (and you cannot sharpen it further), be sure to sew on the *inside* of this marking line. Cut out the pieces, leaving a margin of a little more than a quarter of an inch beyond the seam line.

4. Again following your own colour chart, begin to make up the 9 triangles. Sew *only* from marked corner to corner, *not* through into the seam allowances. Match and pin all the 'matching' points. It will be good to pin between these also. The more pinned points there are the smoother a curve you will achieve. Sew with the 'gathered' side uppermost. Press from concave to convex side.

5. Trim each triangle to give you a 5¼" side length *plus* an exact quarter inch seam allowance on the three outer sides. (This doesn't come to an exact measurement, unfortunately, so it is best to add the quarter with your ruler and rotary cutter as you trim the triangles – by putting the quarter inch line of your ruler along the drawn seam line. It should be just under 6 1/8".)

6. Assemble the completed triangles, according to the diagram, and your own colour chart. Sew them into two rows plus one triangle. Press. Sew the rows together. Press. Set aside.

7. Make templates for the trapezium-shaped divisions of the second set of triangles – *Fig.7,iii (PP)*. Cut these out with a little more than a quarter inch margin.

8. Sew the trapezia together, from marked corner to corner, as your colour chart. Note: you are always sewing a top to a side. To join the third piece to the first two pieces, an inset (or Y) seam is required. I suggest that this is done by sewing the seam in two parts, from the centre to the outside in both cases. It is easier to be accurate this way, rather than starting at the outside, pivoting in the centre and going on to the other side. I find I can get a flatter result, too, with this method. Press all the seams of each triangle in the same direction. The 'ears' at the centre will lie flat, if turned slightly, opened out and pressed. Trim these triangles to 5¼" *plus* an exact quarter inch seam allowance, as before.

9. Make a template for, and cut out 12 of the unpieced, third set of triangles – *Fig.7,iv (PP)*. These can be cut at the exact size, from their template, straight away. These triangles are later incorporated into the side panels – *Fig.7,vii.*

10. Make templates for, and cut out, the background pieces – *Figs.7,v and vi (PP)*.

    You will need 12 of *Fig.7,v*; and 6 of *Fig.7,vi*. If your background fabric has even a hint of a directional pattern, it is a good idea to place the templates so that the grain line of the fabric is going in the same direction on every one. This will give a much better finished appearance.

11. Assemble the side panels, as diagram – *Fig.7,vii.* Press.

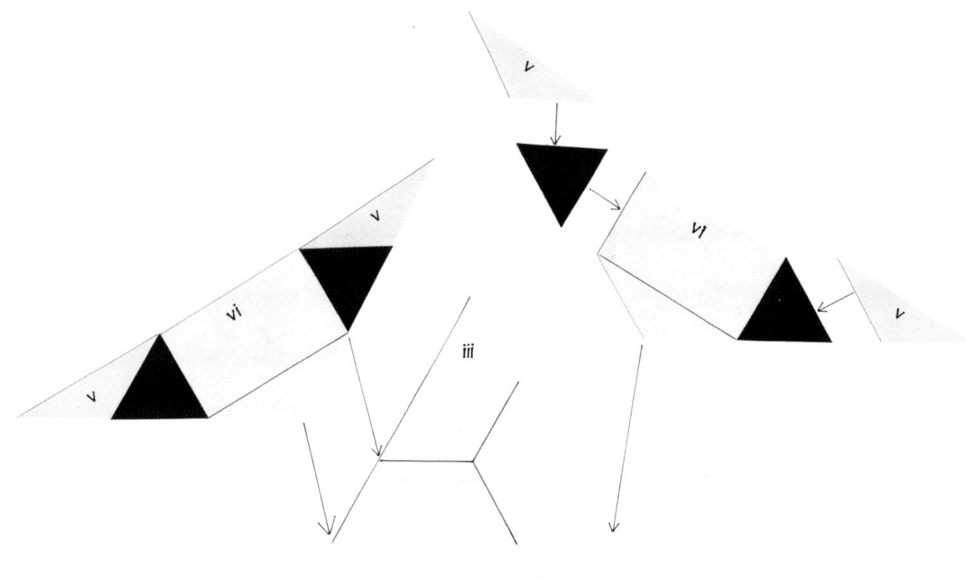


*Fig.7,vii*

12. Attach the side panels to each side of the central triangle. Press towards the central triangle. This creates a regular hexagon.

13. Make templates for and cut out the corner setting triangles – *Fig.7,viii (PP)*, from the background fabric (maintaining the grain line direction that you used for the background pieces of the side panels). You need two triangles from the template as it is, and two triangles from the template reversed. Attach them to the sloping sides of the hexagon. Press towards the hexagon.

14. In order to convert this rectangle to an 18½" square (including seam allowance): cut two strips of background fabric 18½" x 1 5/8", and sew one to each side. Press.

    - If you plan to use this centre piece as a block in a large quilt, such as those described in Section 3, your work is complete.

- If, however you are making a wallhanging or similar small quilt, finishing instructions – common to all designs – are given at the end of this section.
- Possible quilting designs are given below -– *Figs.7,ix–xi.*
- See also full quilting details at the end of this Section.

## Cornerstones

These cornerstones are not made by Foundation paper piecing. They use a technique of folding on the bias grain, rolling back this edge and stitching it back to make a curve.

The four corners each comprise four squares, made in the same way; but the curves are placed in four different ways.

Each square requires two 1½" squares of contrasting fabric. I used fabric of the central design, on the background.

Fold the piece which will be on top along its diagonal and pin (or tack/baste) each of its corners to one half of the background piece to secure it. Roll back the bias edge, and stitch down. You will need to make 16 altogether.

After you have decided on your placement, sew four squares together for each cornerstone.

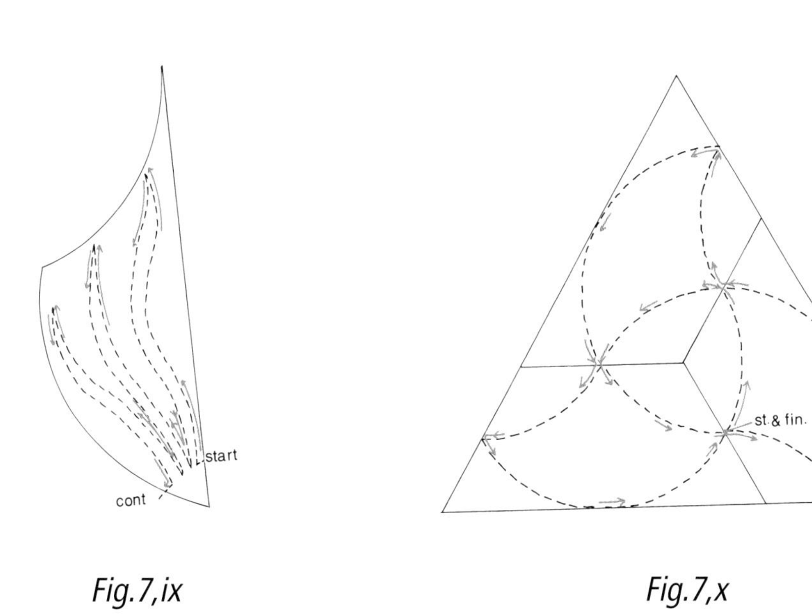


*Fig.7,ix*

*Fig.7,x*

*Fig.7,xi*

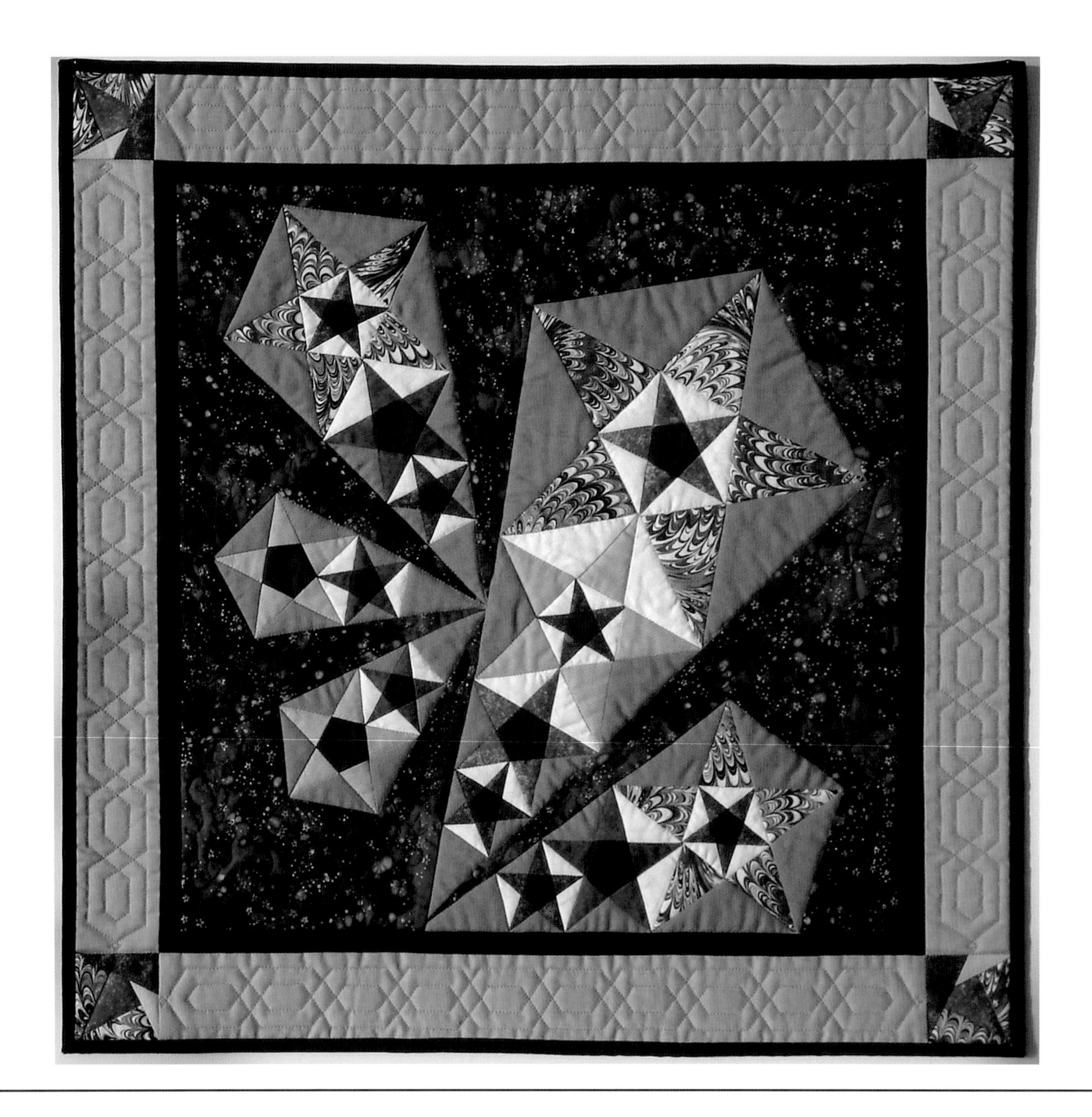

# Chapter 8 – LUTES OF PYTHAGORAS

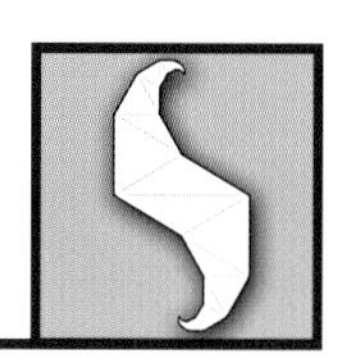

*The Lute of Pythagoras is a fractal. The shape is based on the Golden Triangle, where the ratio of base to sides is 1:1.618. The flat base of the Golden Triangle is altered so that the base angles become those of a regular pentagon. The Triangle – or, now, Lute – may be divided into successive, interlocking pentagons, to infinity; and each pentagon may have pentagrams drawn within it, also to infinity: although, practically, there is always a stopping point.*

## Fabric Selection

- Fabric 1, for the outer edge pieces of all the Lutes.
- Fabric 2, A – D or more. Different fabrics, which contrast well with one another. These may be repeated.
- Fabric 3. Background fabric which should offer some contrast, particularly with Fabric 1. The background piece needs to be about 1" larger, all round, than the size you want it to be when finished.

**Plus:** *if you are making a complete, small quilt:*

- Fabric 4. For narrow border and binding. (Cut 1" and 1¼" wide strips, respectively)
- Fabric 5. For wider border. This may be the same as Fabric 3, if you like the look of the background extending beyond the narrow border. (Cut 2½" wide strips)
- Small pieces of fabrics used elsewhere in the quilt, for the pieced cornerstones.
- Backing fabric. A piece slightly larger all round than the finished quilt will be.

## Assembly Instructions

**1.** You will see from the Pattern for the Lute, that there are several, potential 'cut-off' points. You may choose your own sizes and placement for the Lutes. If you do so, draw a sketch of your design and colour it to show which fabric you will place where. A basic design, to use as a colour chart is given – *Fig.8,i.*

*(If you wish to make a larger Lute than the one given here, use the baseline of the given Golden Triangle as the length of the side of the next regular pentagon, and draw in the pentagram to match those shown).*

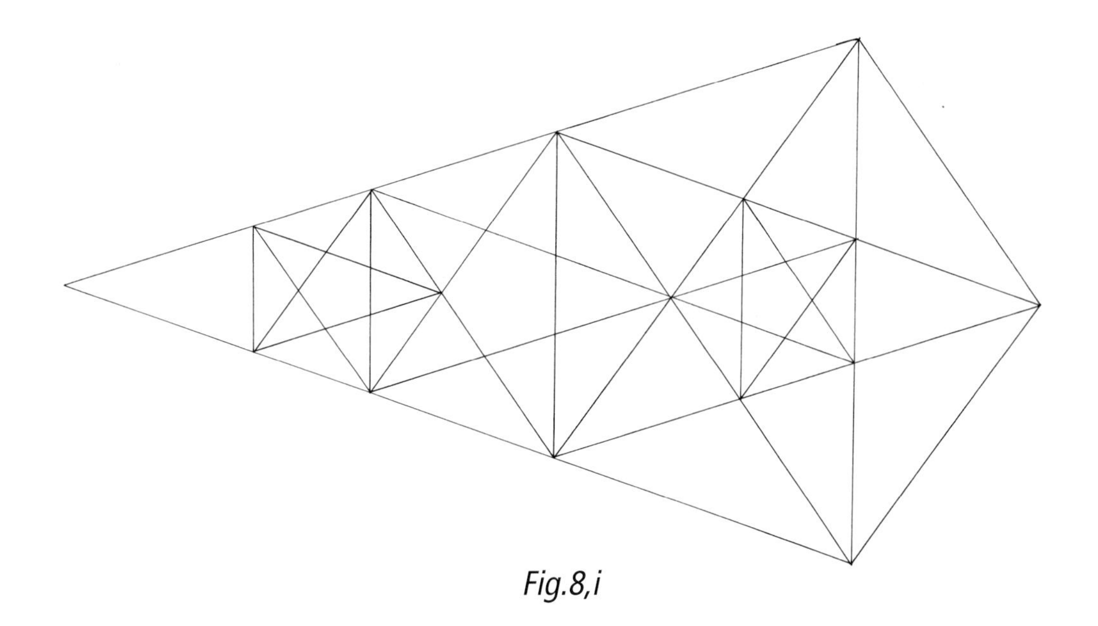

*Fig.8,i*

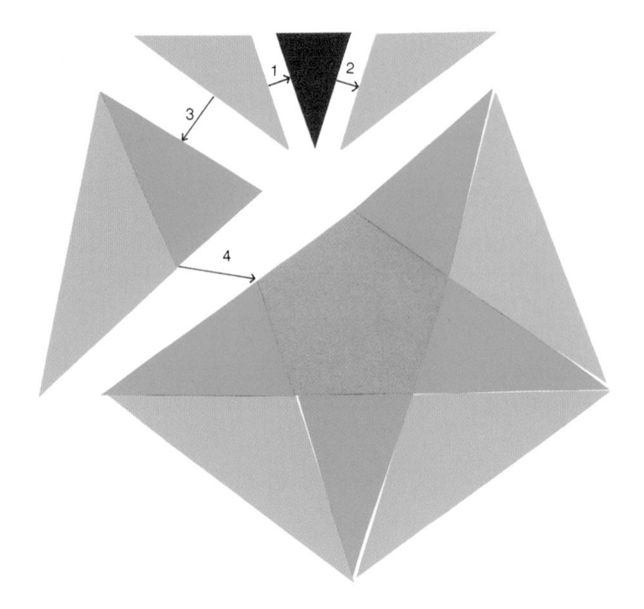


*Fig.8,iii*

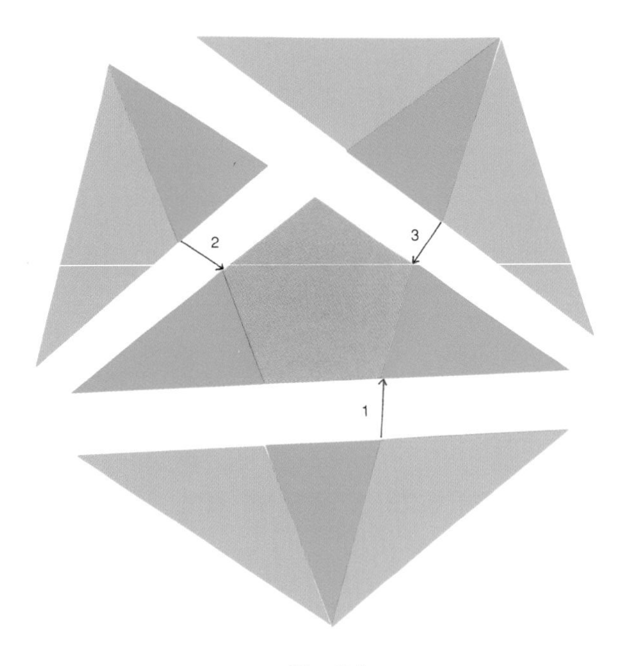


*Fig.8,iv*

2. From the basic Lute pattern – *Fig.8,ii (PP)*, trace as many Lutes as you wish to include in your design (if different from the example), in the sizes you require, onto freezer paper. Before cutting these apart, number each Lute and all parts of it; plus a mark to indicate which fabric you propose to use for each piece.

3. Assemble your fabrics and press the appropriate, freezer paper pattern pieces to the wrong side of their fabrics. Note that, depending on where each star is placed, it will or will not require a pieced outer triangle.

4. Cut out all these pieces with approximately a quarter inch seam allowance. Put the pieces together according to the marking system you employed at the start.

5. Sew together, according to the Assembly Diagrams – *Fig.8,iii (pentagram with one pieced outer triangle) and iv (basic pentagram*

*without a pieced element).* There will be no inset seams if you follow the piecing sequences shown. Press.

**6.** Join the sections of each Lute together, matching seams carefully. Press.

**7.** Trim around each Lute, leaving an allowance of about a quarter of an inch. Fold these allowances over the edges of the paper templates and press them under. Remove the papers.

**8.** Appliqué the Lutes to the background in whichever arrangement you have decided upon. Press. Trim and square this centre piece to the required size.

- If you plan to use the centre piece as a block in a large quilt, such as those described in Section 3, the work for this design is now complete.
- If, however, you are making a wallhanging, or similar small quilt, the finishing instructions, which are common to all the designs, are given at the end of this section.
- Possible quilting designs are given below – *Figs.8,v and vi.*
- See also full quilting details at the end of this section.

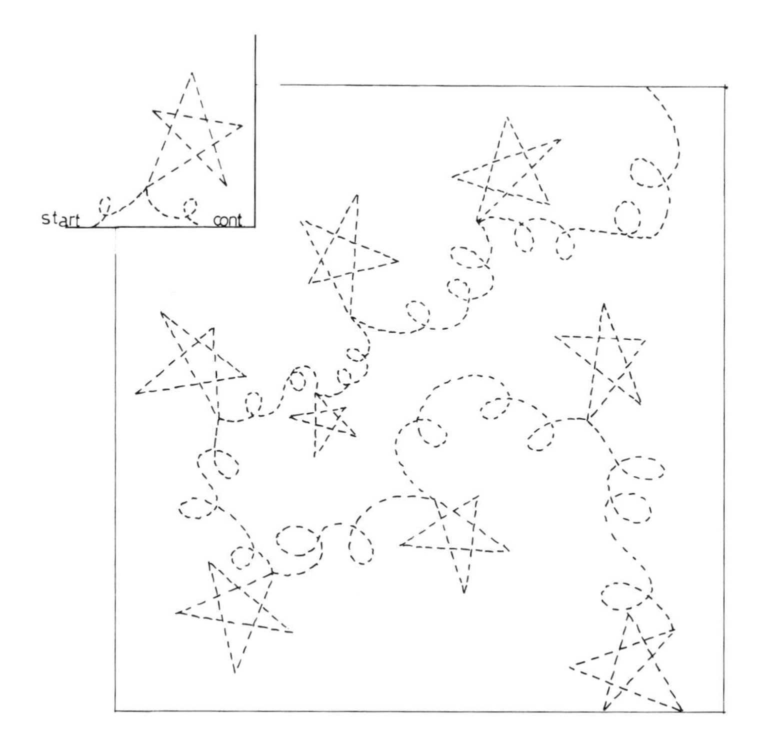


*Fig.8,v*

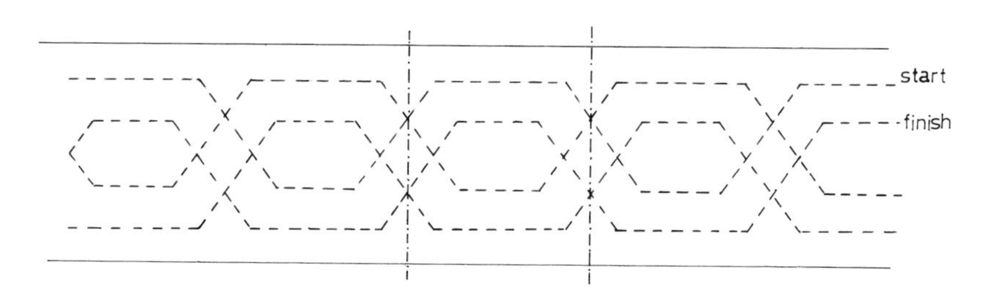


*Fig.8,vi*

# Chapter 9 – ROTATIONAL SYMMETRY

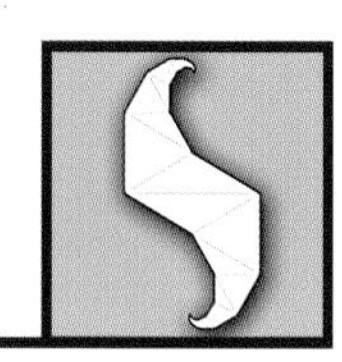

*An image is said to have rotational symmetry if there is a centre point around which the object is turned a number of times through a certain number of degrees and the object remains unchanged in itself. The sub-divided wedge shape in this small quilt has been turned ten times, through 36° each time.*

## Fabric Selection

- Two basic colours which have some contrast between them (here, green and blue). The interlocking effect in this design depends on having a light and a dark version of both of these colours. Fabrics 1A (light/dark), 1B (dark/light), 2A (light/dark) and 2B (dark/light).
- Fabric 3. A third different colour, common to all wedges, as the outer piece.
- Fabric 4. Background.

**Plus:** *if you are making the complete, small quilt:*

- Fabric 5. For the narrow border and binding. (Cut 1" and 1¼" wide strips, respectively.)
- Fabric 6. For the wider border. This can be the same as Fabric 4, if you like the idea of the background extending beyond the narrow border. (Cut 2½" wide strips.)
- Small pieces of fabrics used elsewhere in the quilt, for the pieced cornerstones.
- Backing fabric. A piece slightly larger all round than the finished quilt will be.

## Assembly Instructions

**1.** Trace ten copies of the wedge in the Pattern section onto paper suitable for Foundation Paper Piecing – *Fig.9,i (PP)*. Five of the wedges will be based on Fabric 1; the other five will be based on Fabric 2.

*(It would be quite straightforward to piece this design using simple templates, if you prefer.)*

**2.** Before piecing, I suggest marking the divisions of the wedges in some way which will clearly identify the fabrics you will be using. The interlocking effect is achieved by repeating, in each wedge, the fabric used in the previous wedge adjacent to piece 2 in the sequence.

3. Once the piecing is complete, trim all the wedges so that they have an accurate quarter inch seam. Remove papers. At the apex of the triangle, for each wedge, draw in the seam lines. Press.

4. The points where the seam lines intersect are at the exact centre of the design.

5. Sew the wedges together, in pairs, alternating the base colours – *Fig.9,ii.* Match the centre points very carefully. Begin sewing the seams from the centre and sew out towards the edges. Join the pairs together, continuing to match the centre points carefully. Press.

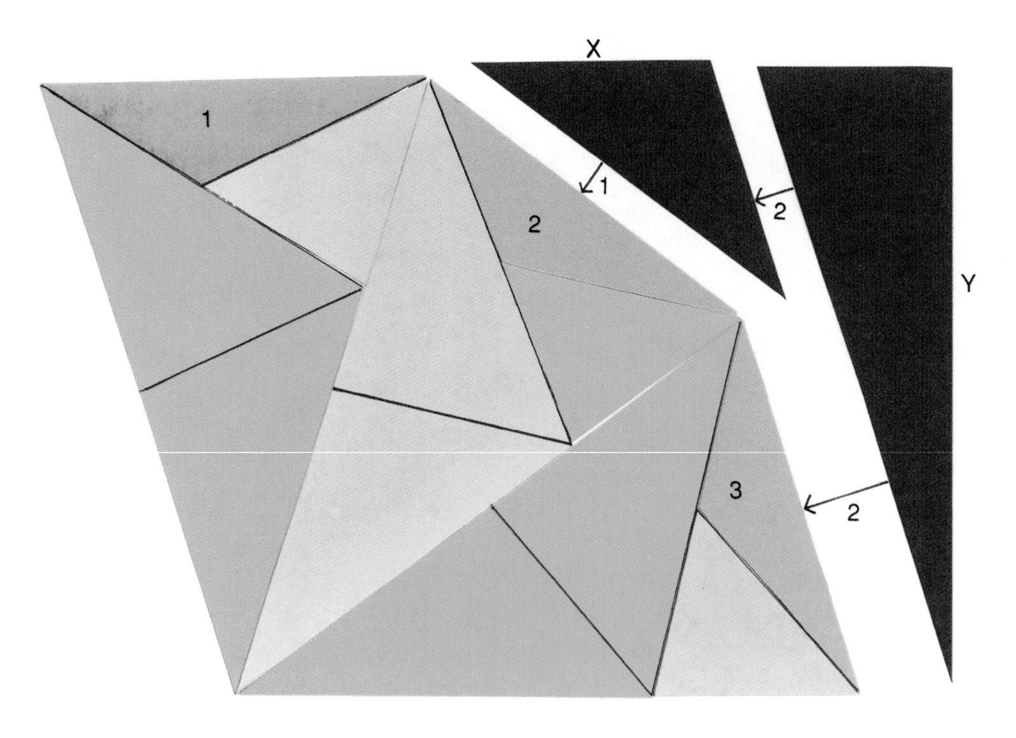


*Fig.9,ii*

6. Make templates for the background pieces, using the patterns in the Pattern section. Cut four pieces of Pattern X – *Fig.9,iii (PP).* As Pattern Y – *Fig.9,iv (PP)* is not symmetrical, two pieces will have to be cut as the pattern and two with the template reversed.

7. If the sides of the final shape are numbered, notionally, clockwise from the top, the X pattern pieces should be sewn to sides 2, 5, 7 and 10. Press.

8. The Y pattern pieces should be sewn to sides 3, and 8; The Y reverse pattern pieces should be sewn to sides 4 and 9 (plus the adjacent side of the X piece in all cases). This produces the final rectangular shape. Press.

   - If you plan to use the centre piece as a block in a large quilt, such as those described in Section 3, the work for this design is now complete.

   - If, however, you are making a wallhanging, or similar small quilt, the finishing instructions, which are common to all the designs, are given at the end of this section.

   - Possible quilting designs are given below – *Figs.9,v–vii.*

   - See also full quilting details at the end of this section.

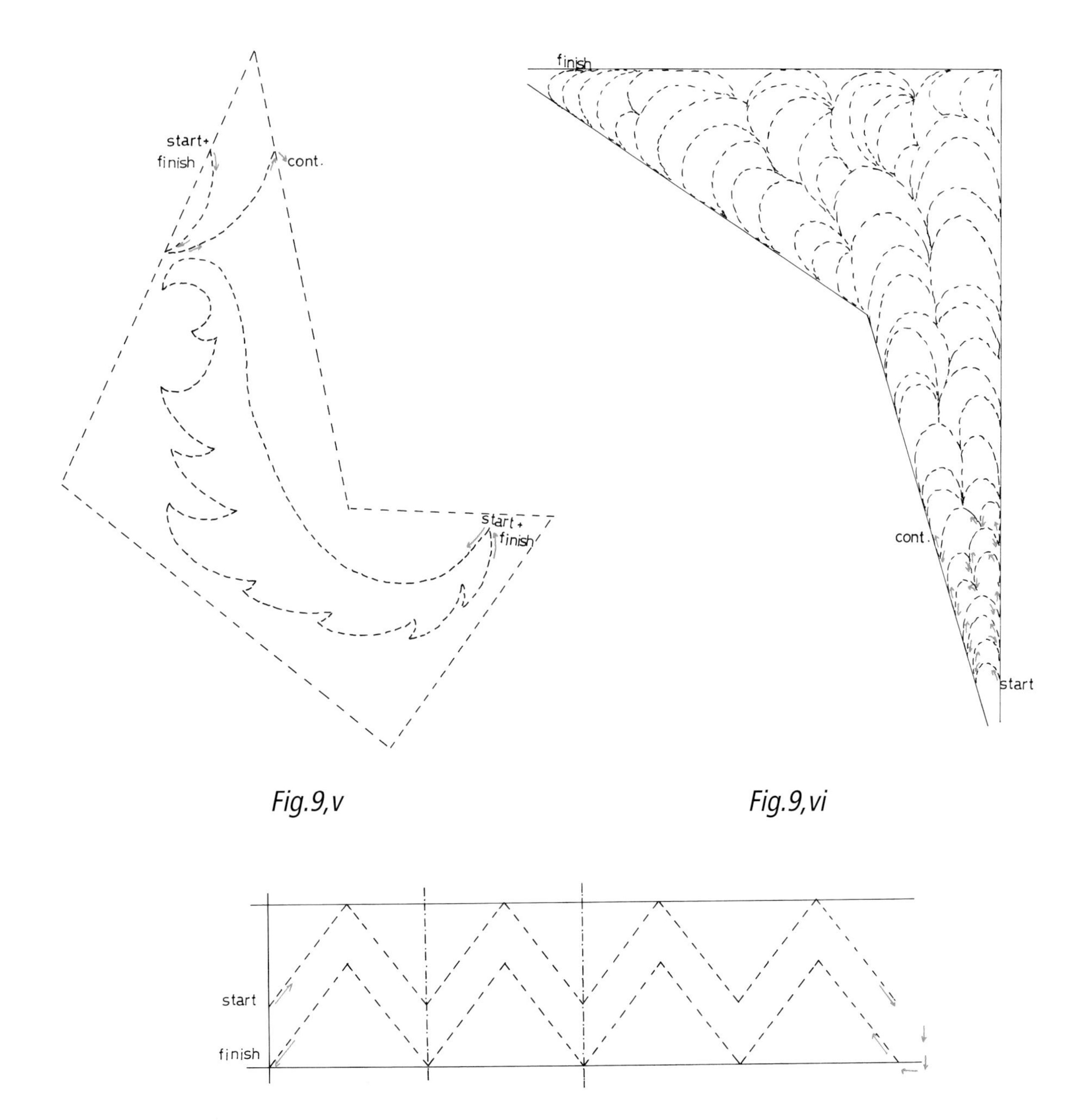


*Fig.9,v*

*Fig.9,vi*

*Fig.9,vii*

# Chapter 10 – REFLECTIVE SYMMETRY

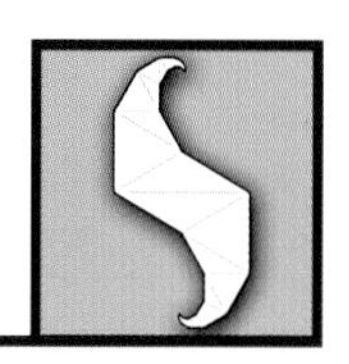

*Reflective Symmetry is a line symmetry. A shape is said to have reflective symmetry if the shapes on either side of the axis (or axes) of symmetry are the same shape, though one is the mirror image of the other. This dodecagon has six axes of symmetry and includes six pairs of reflectively symmetrical wedges.*

## Fabric Selection

- Fabrics for the central shape itself. In my example I have used a different fabric for each piece in every wedge – 11 in all. These are not completely unrelated, however. They fall into two groups: blue/mauve and yellow/orange. I based my colour choice on the fabric I had already chosen for the wider border. Fabrics 1A - 1K.

- Fabric 2. For the background.

**Plus:** *if you are making a complete, small quilt:*

- Fabric 3. For the narrow border and binding. (Cut 1" and 1¼" wide strips, respectively.)

- Fabric 4. For the wider border. This can be the same as Fabric 2, if you like the look of the background extending beyond the narrow border. (Cut 2½" wide strips.)

- Small pieces of fabrics used elsewhere in the quilt, for the pieced cornerstones.

- Backing fabric. A piece slightly larger all round than the finished quilt will be.

## Assembly Instructions

1. Trace 6 copies of Pattern A – *Fig.10,i (PP)* and 6 copies of Pattern B – *Fig.10,ii (PP)*, (the mirror image) to paper suitable for Foundation Paper Piecing. Number these wedges, clockwise, from the top. Colour or number the pieces to identify clearly which fabric will be going where.

2. Piece all twelve wedges. Trim to shape and size adding an exact quarter inch seam allowance. Draw the seamlines at each central point for the most successful centre joins. Press. Remove paper.

3. Join the wedges in pairs, matching seams. Begin sewing from the centre out towards to the edges. Join three pairs of wedges to form

one half, and join the remaining three pairs to form the second half. Finally, sew the two halves together, matching centre points and seams precisely. Press.

4. Make templates from the remaining three pattern pieces – *Figs.10,iii–v (PP)*, to create the background. Cut four pieces of Pattern C; cut two pieces of Patterns D and E; cut two pieces of Patterns D and E, reversed – all from background fabric.

5. See Fig.10,vi. Attach a background piece C to sides 2, 5, 8 and 11.

   Attach a background piece D to sides 1 and 7, plus the adjacent side of the previous piece C.

   Attach a background piece reverse D (DR) to sides 6 and 12, plus the adjacent side of the previous piece C.

   Attach a background piece E to sides 3 and 9, plus the free ends of the previous D and C pieces.

   Finally, attach a background piece reverse E (ER) to sides 4 and 10, plus the free ends of the previous DR and C pieces. Press.

6. Trim and square the quilt so far.

   - If you plan to use the centre piece as a block in a large quilt, such as those described in Section 3, the work for this design is now complete.

   - If, however, you are making a wallhanging, or similar small quilt, the finishing instructions, which are common to all designs, are given at the end of this section.

- Possible quilting designs are given below – *Figs.10,vii–ix*.

- See also full quilting details at the end of this Section.

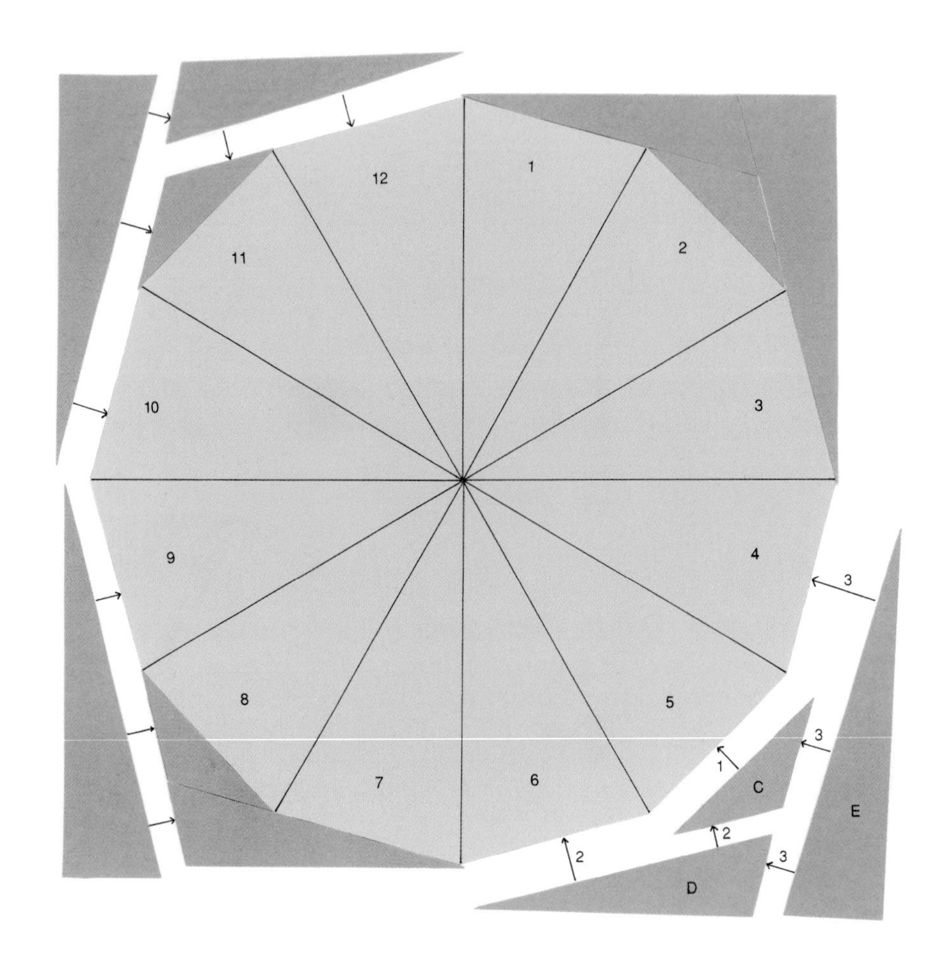


*Fig.10,vi*

*Fig.10,vii*

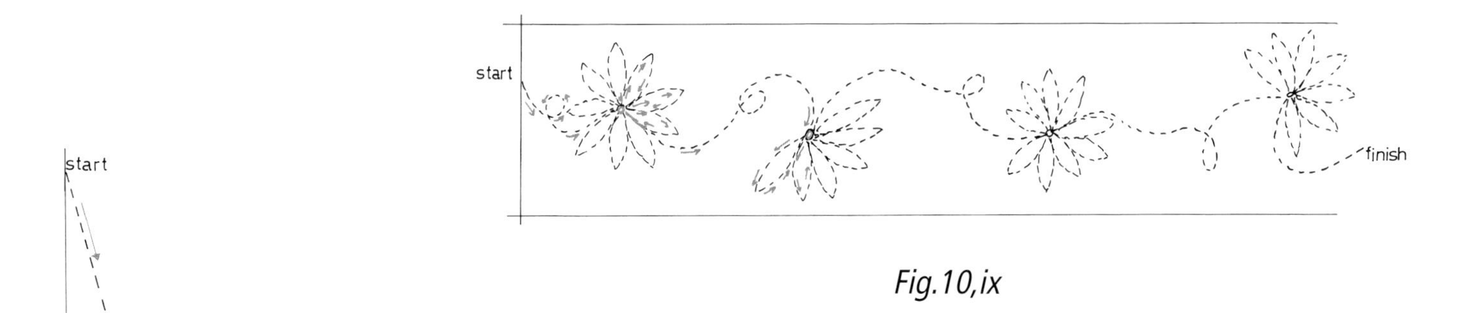


*Fig.10,ix*

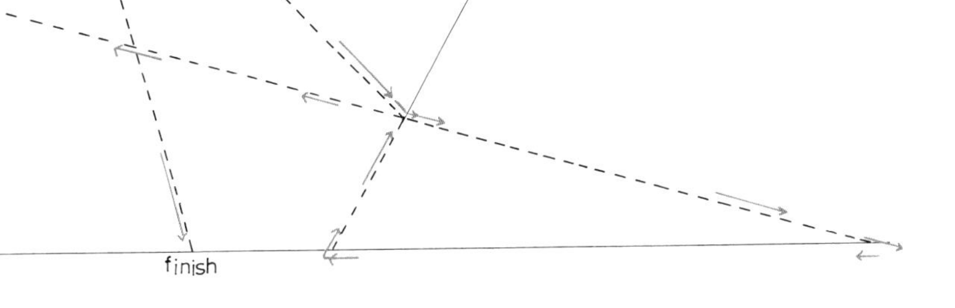


*Fig.10,viii*

# Chapter 11 – PENROSE TILING

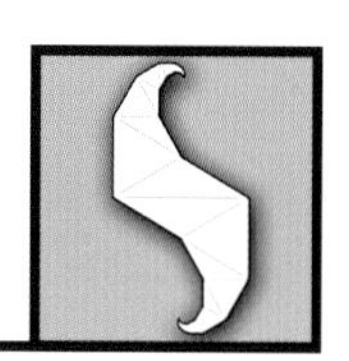

*A method of using two differently shaped diamonds (36° and 72°) in order to make tessellating shapes based on a regular pentagon or decagon, which do not otherwise tessellate. The tiling pattern shown here is based on decagonal shapes.*

## Fabric Selection

- At least two colours of good contrast for the diamond design. It is a good idea to make a drawing of the pattern – rough will do – and colour this so that you know where each diamond will go. Fabrics 1A, 1B, 1C...

  My example quilt uses just two colours for the centre piece and background. The variation below shows the wonderful effects that can be achieved using several more colours.

- Fabric 2. A piece of background fabric, about 1½" larger all round than the size you want it to be when finished.

**Plus:** *if you are making a complete, small quilt:*

- Fabric 3. For the narrow border and the binding. (Cut 1" and 1¼" wide strips, respectively.)
- Fabric 4. For the wider border. This could be the same as Fabric 2, if you like the effect of the same fabric running 'under' the narrow border. (Cut 2½" wide strips.)
- Small pieces of fabrics used elsewhere in the quilt, for the pieced cornerstones.
- Backing fabric. A piece slightly larger all round than the finished quilt will be.

## Assembly Instructions

1. Using the diamonds in the Pattern Section – *Fig.11,i (PP)*, make a template of each (on fine card or template plastic) and use these to draw the diamonds onto your chosen base paper. I usually use large, white business envelopes. When drawing round the templates, use a sharp pencil and angle the point into the join between the template and base paper.
2. Cut 50 x 36° diamonds and 70 x 72° diamonds from the base paper.

3. Having chosen your fabrics, cut one piece of fabric for each piece of base paper. The fabric should be cut, using the templates, allowing at least an extra quarter of an inch all round. This does not need to be cut accurately.

4. Baste/tack your fabric pieces to the base paper. My preferred method is to place the fabric right side up onto the base paper, and fold down along all four sides, finger pressing to get a sharp crease. Then baste/tack the folded-under fabric to the top, sewing through the base paper – *Fig.11,ii.* No need for small neat stitches at this stage. Baste/tack along all four sides. The large angle corners will turn under completely, the small angle corners will have 'ears' sticking out. Don't attempt to turn these under, they will disappear at the next stage without that. Press all pieces.

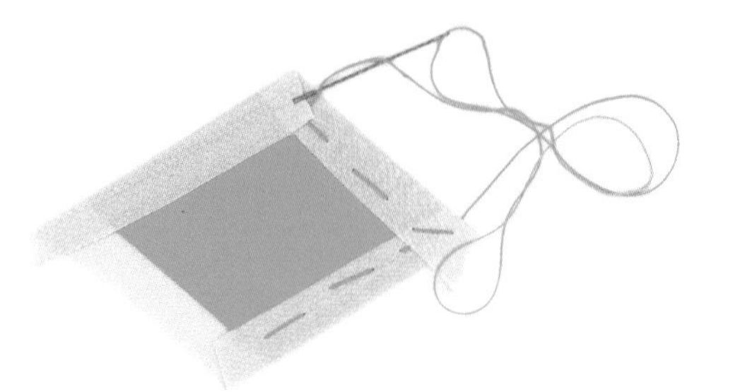

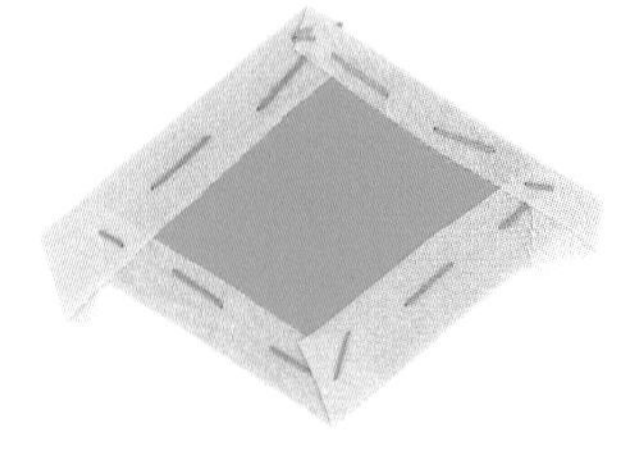

*Fig.11,ii*

5. When all your pieces are ready, assemble the design on a table, piece of board or design wall, according to the colour placement you have decided upon. You might need to alter your choice of position at this stage. This 'rehearsal' is particularly important if you are using more than two colours.

6. Beginning at the centre, then proceeding systematically out from the first 'round', the pieces can now be sewn together. See Assembly Detail below – *Fig.11,iii* and also *Fig.11,iv*. Putting right sides together, join as required from the back, using a very small, oversewing or whip stitch. Using a fine silk thread, matching or merging with the colours you are using, makes these stitches almost invisible. Make sure you keep the surplus fabric of the small corners on the side facing you – the back. Press as you go.

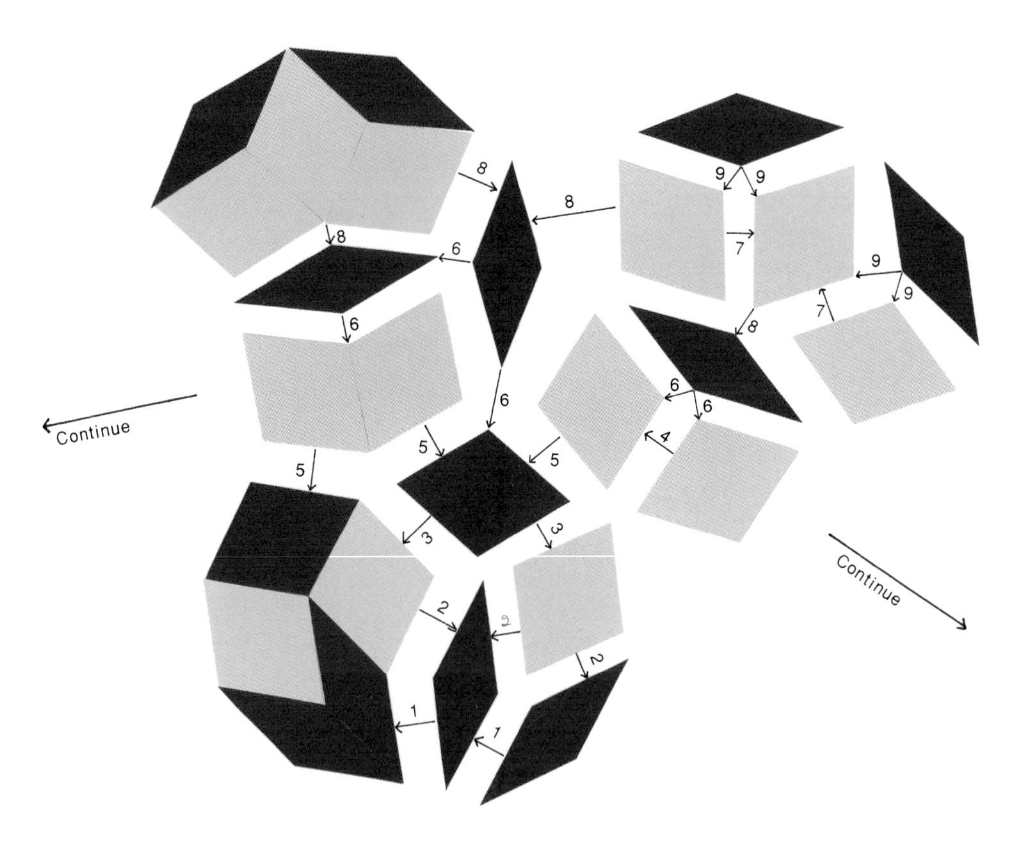


*Fig.11,iii*

*Fig.11,iv*

When all the diamonds have been joined together, press flat. Then appliqué the piece to your chosen background. Try to keep the outside angles - inward and outward - as sharp as possible. Press.

**7.** Trim this centre piece to the desired size.

- If you plan to use this centre piece as a block in a large quilt, such as those described in Section 3, the work for this design is now complete.
- If, however, you are making a wallhanging, or similar small quilt, the finishing instructions, which are common to all the designs, are given at the end of this section.
- Possible quilting designs are given below – *Fig.11,v and vi.*
- See also full quilting details at the end of this section.

## Assembly Detail

The first round consists of 10 narrow diamonds, sewn with the small angle towards the centre.

Then a round of the wide diamonds with the small angle set into the gaps of the first round.

The third round is made up of wide diamonds, with the large angle set into the gaps of round two.

For the fourth round, first join pairs of wide diamonds with their small angles 'downwards'. Set these pairs into the gaps of the previous round.

The fifth round consists of narrow diamonds, set into the gaps of the previous round, but this time they are alternately set with small and large angle to fit the gap.

For the sixth round, make ten units of three wide diamonds. These are set to fit between the high and low peaks of the previous round.

The final round consists of narrow diamonds, simply set to fill the gaps in the previous round.

This results in a concave edge pattern. A further round of wide diamonds could easily be set in to give the appearance of an almost circular motif.

*Sarah's Alternative Version*

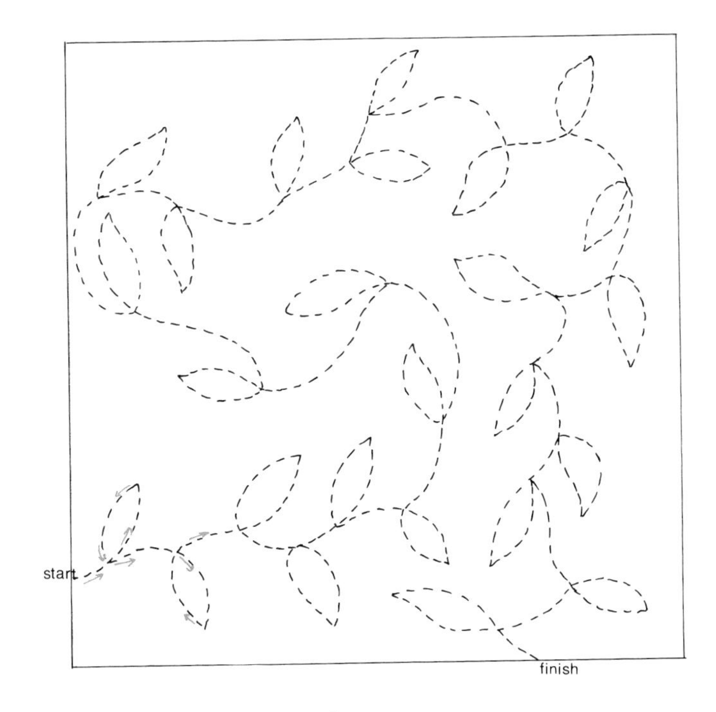


*Fig.11,v*

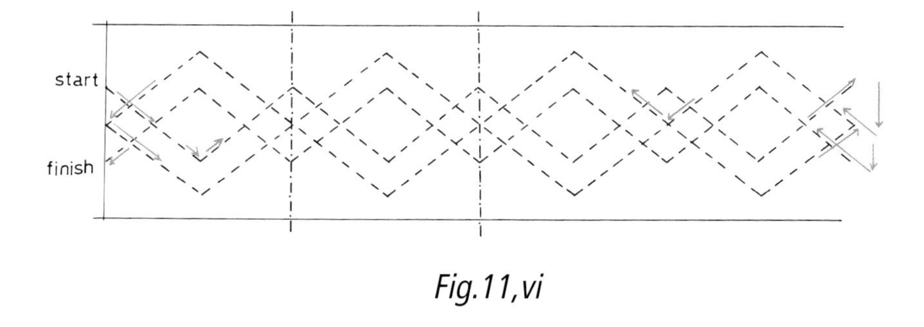


*Fig.11,vi*

# Chapter 12 – TESSELLATION

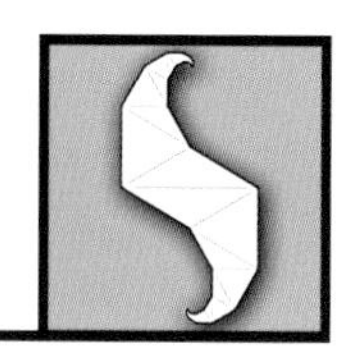

*A simple tessellation may be defined as the covering of an area with regular shapes which tessellate together with no gaps. Some regular shapes do tessellate, for example: isosceles and equilateral triangles, squares, rectangles, hexagons, rhombi and parallelograms. Some trapezia will tessellate. Familiar shapes which will not tessellate are: octagons, pentagons and circles (or any part thereof).*

## Fabric Selection

- Two fabrics of good contrast for the centre design. Fabrics 1A and 1B.

  Do not feel restricted to just two fabrics. An example of the quilt made with very many different fabrics (scrappy quilt in pinks and yellows) is shown below.

**Plus:** *if you are making the complete, small quilt:*

- Fabric 2. For the narrow border and binding. (Cut 1″ and 1¼″ wide strips, respectively)
- Fabric 3. For the wider border. (Cut 2½″ wide strips)
- Small pieces of fabrics used elsewhere in the quilt, for the pieced cornerstones.
- Backing fabric. A piece slightly larger all round than the finished quilt will be.

## Assembly Instructions

1. Cut several 1½″ wide strips from Fabrics 1A and 1B (perhaps start with five, if using fat quarters or similar).
2. Taking three of each colour, make two sets of joined strips, as follows, and – *Fig.12,i*:

   **i.** Fabric 1A, Fabric 1B, Fabric 1A

   **ii.** Fabric 1B, Fabric 1A, Fabric 1B
3. Cut each of these strip sets into 18 x 1½″ wide pieces.
4. Cut the remaining strips into 3½″ pieces. You need 36 of each colour. Put these strip pieces, together with the strip set pieces, so as to make Blocks A and B, according to the Assembly diagram. 18 of each block are required to complete the centre square. Press.

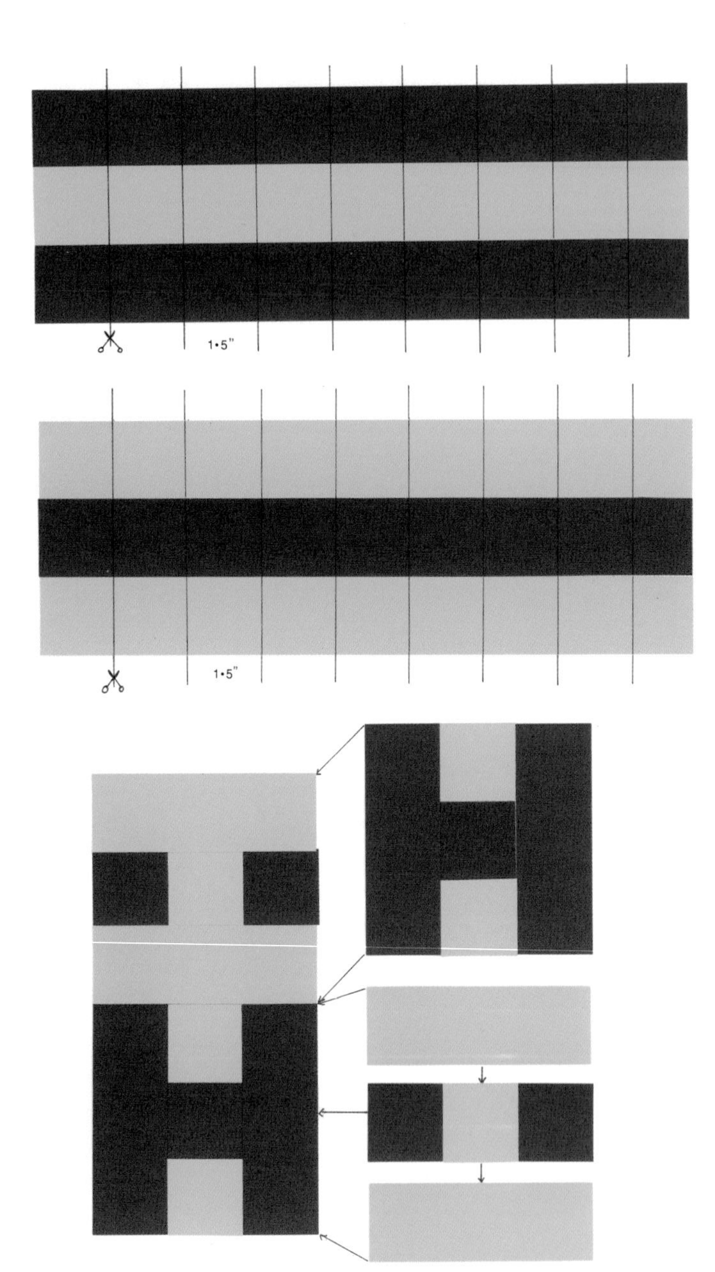


*Fig.12,i*

This procedure of cutting strips might have to be repeated, as the number of pieces obtained from each strip or strip set depends on the size of the piece of fabric used, and on the way the strips are cut.

5. Join the blocks in rows of six, alternating the main colours and the orientation of the 'H'. Press.

6. Join the six rows to complete the square. Join first in pairs, and then join the pairs of rows together. Press.

   - If you plan to use this centre piece as a block in a large quilt, such as those described in Section 3, your work is now complete.
   - If, however, you are making a wallhanging or similar small quilt, the finishing instructions, which are common to all the designs, are given at the end of this section.
   - Possible quilting designs are given below – Figs.12,ii–iv.
   - See also full quilting details at the end of this Section.

*Helen's Alternative Version*

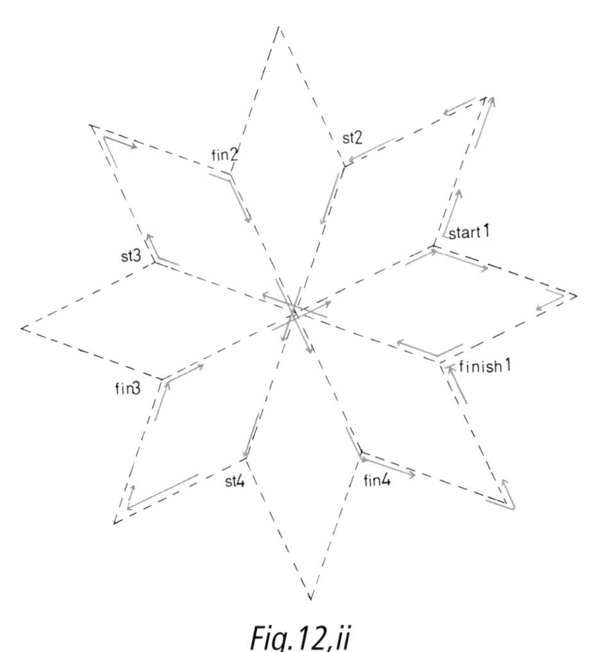


*Fig.12,ii*

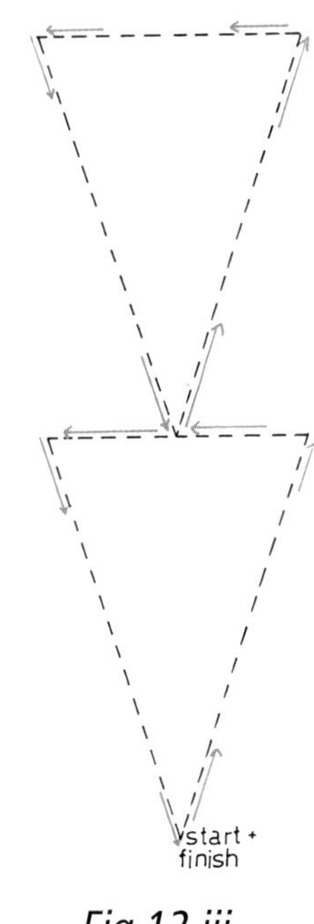


*Fig.12,iii*

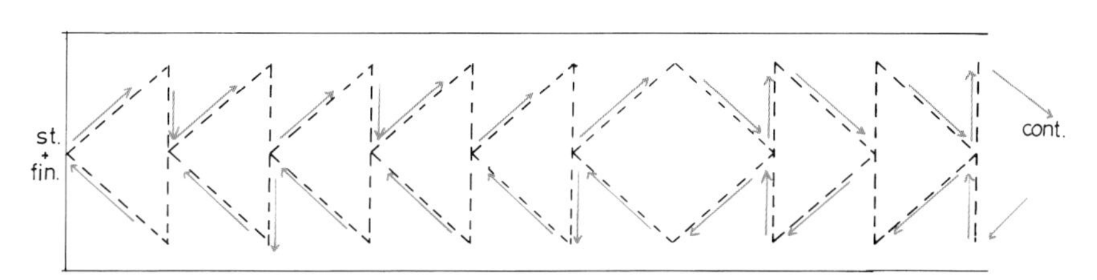


*Fig.12,iv*

# Chapter 13 – FIBONACCI SERIES

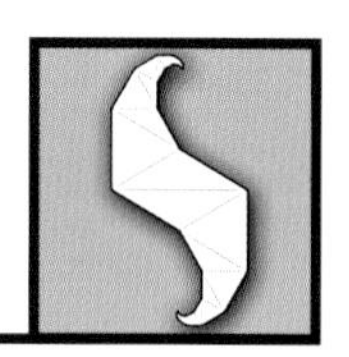

*A numerical sequence, starting from 1, where each number in the sequence is the sum of the two previous numbers. The series was discovered by Leonardo Fibonacci near the turn of the 12th/13th century. It is used a lot in design because the relationship between the elements in the series can easily be translated into spatial ratios.*

*1 {1, 2, 3, 5, 8, 13, 21, 34} 55, 89, 144 ... and so on. { } used in this design.*

## Fabric Selection

- Eight fabrics for the measured strips. These may all be different, or some may be repeated, as in my example. Fabrics 1A, 1B, 1C and so on. Fabrics should be chosen so that there is a definite distinction between strips – colour contrast or dark/light.

**Plus:** *if you are making the complete small quilt:*

- Fabric 2. For the narrow border and binding. (Cut 1" and 1¼" wide strips, respectively.)
- Fabric 3. For the wider border. As the width of this strip is in no way connected with the series, the fabric should not be the same as any of the Fabric 1 strips. (Cut 2 ½" wide strips.)
- Small pieces of fabrics used elsewhere in the quilt, for the pieced cornerstones.
- Backing fabric. A piece slightly larger all round than the finished quilt will be.

## Assembly Instructions (Simple Courthouse Steps Pattern)

**1.** I suggest using the Foundation Paper Piecing pattern in the Pattern Section to make the centre – *Fig.13,i (PP).*

It is possible to make this first part of the quilt by cutting strips of ¾", ¾", 1", 1¼" and 1¾", and building these up on a 1/2" (plus seam allowances) square centre; but the Foundation Paper Piecing method will be more accurate, quicker and easier. Trim on the dotted line, press and remove papers.

**2.** Now cut strips to complete the square:

2 of 2½" x 7" (seventh colour); 2 of 2½" x 11" (seventh colour); 2 of 3¾" x 11" (eighth colour); 2 of 3¾" x 17½"(eighth colour)

**3.** Maintaining the Courthouse Steps pattern begin with the Foundation pieced centre – *Fig.13,ii* add the 7" strips to the sides. Press. Add the 11" strips (seventh colour) to the top and bottom. Press.

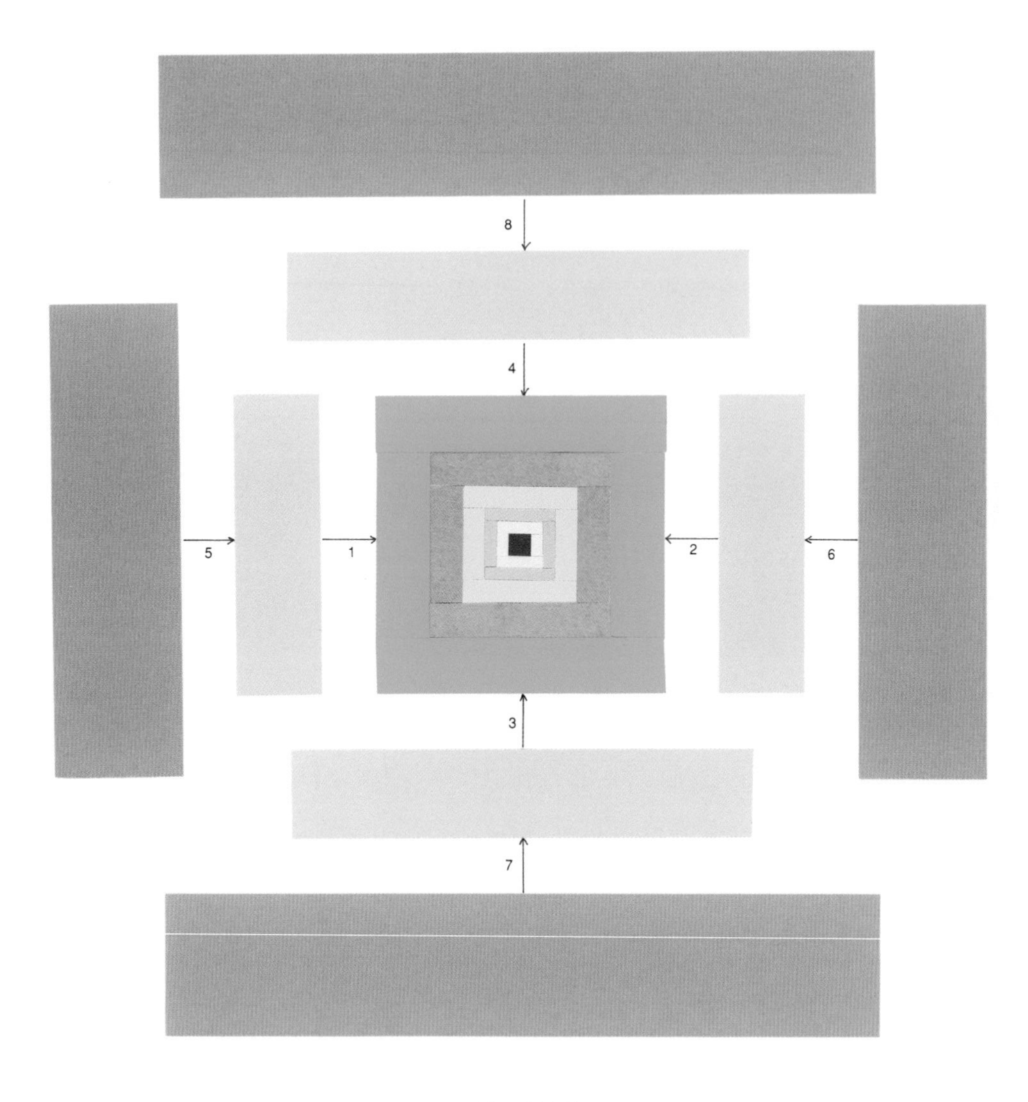


*Fig.13,ii*

**4.** Continuing in the same pattern, add the final strips to make the square: add the 11" strips to the sides. Press. Add the 17½" strips to the top and bottom. Press.

- If you plan to use this centre piece as a block in a large quilt, such as those described in Section 3, your work is now complete.
- If, however, you are making a wallhanging or similar small quilt, the finishing instructions, which are common to all the designs, are given at the end of this Section.
- Possible quilting designs are given below – *Figs.13,iii and iv.*
- See also full quilting details at the end of this Section.

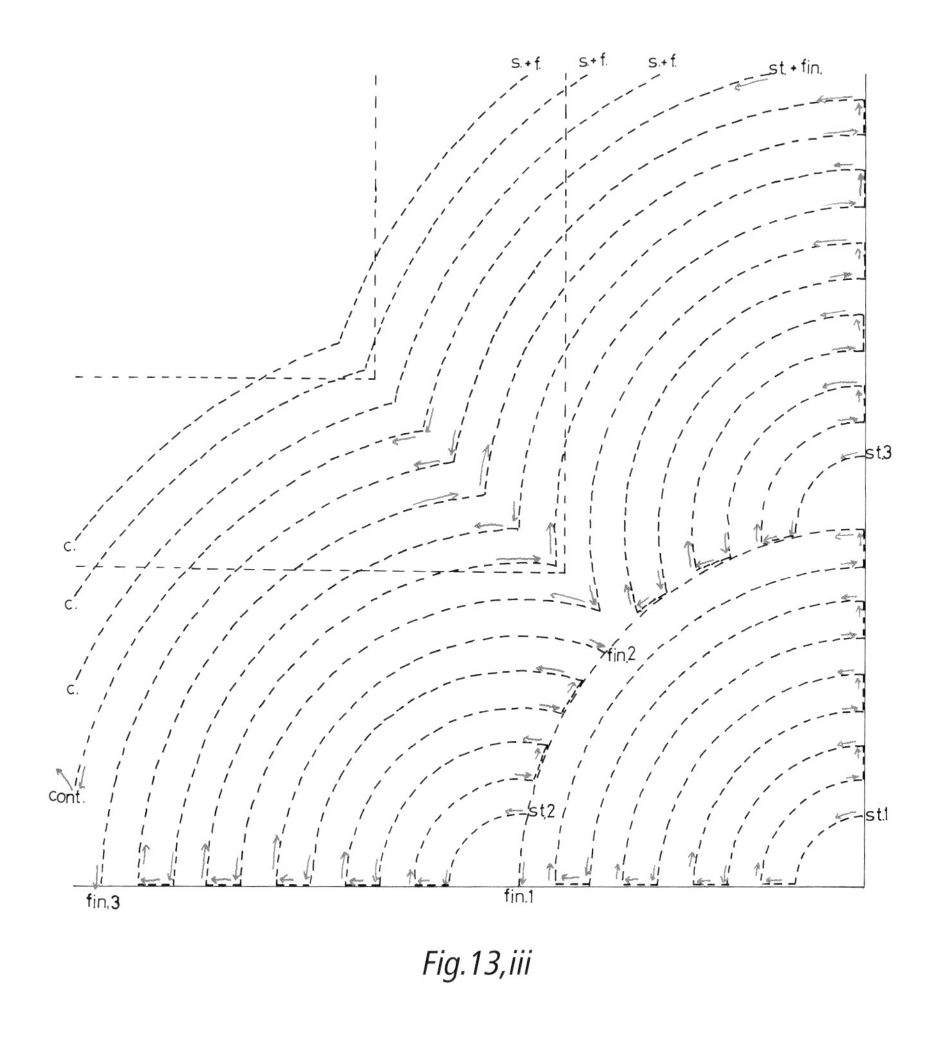


*Fig. 13, iii*

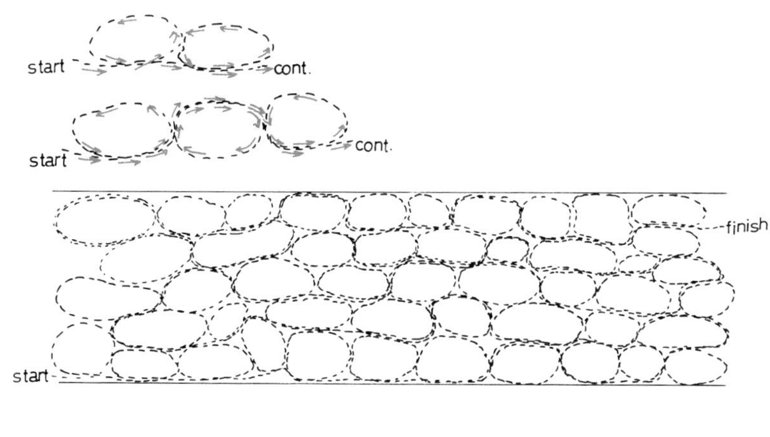


*Fig. 13, iv*

# Chapter 14 – PERSPECTIVE AND OPTICAL ILLUSION

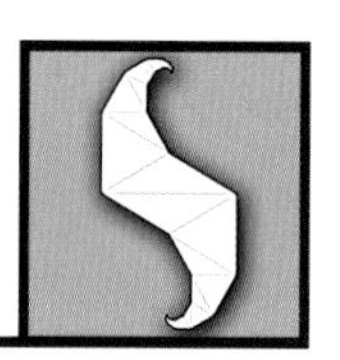

*Visual perspective is the way objects appear to the eye. Linear perspective is a mathematical system for creating the illusion of space, shape and distance (3 dimensions) on a flat surface. The Icosahedron is a 20-faced polyhedron. I have used shape and colour to achieve the effect of a three dimensional shape in this small quilt.*

## Fabric Selection

- Fabric 1 A-D or E. For the Icosahedron. 5 similar fabrics, graded in colour from light to dark, will give the best result. 4 fabrics will give an acceptable effect.
- Fabric 2. For the background to the Icosahedron. A contrasting colour is best for this.
- Fabric 3 A-C. For the surround to this centre square. 3 similar fabrics, graded in colour – light, medium and dark, to create the illusion of a frame with depth.
- Fabric 4. For the inner border of the central framed section.
- Fabric 5. For the background to the stars and flying geese.
- Fabric 6 A - ? For the stars and flying geese.

**Plus:** *if you are making the small quilt or wallhanging:*

- Fabric 7. For the second narrow border and binding. (Cut 1" and 1¼" strips respectively).
- Fabric 8. For the wider border. (Cut 2½" wide strip). This can be the same as Fabric 5, if you like the look of the background extending beyond the narrow border.
- Small pieces of fabrics already used in the quilt, for the cornerstones.
- Backing fabric. A piece slightly larger all round than the finished quilt will be.

## Assembly Instructions

**1.** Trace the Icosahedron from the pattern – *Fig.14,i (PP)*, including the numbers for each piece. Before you cut anything out , write onto each piece which fabric will be used for it and use an arrow to indicate which way up it will be placed. Several pieces are similar in two orientations.

As the Icosahedron is made by the English Paper Piecing method, your tracing needs to be made on a substantial tracing medium, or it will have to be transferred to a heavier weight of paper. I transferred mine using carbon paper.

2. Place each piece on the wrong side of its chosen fabric and cut around it with a sufficient seam allowance for this to be tacked to the paper piece. The seam allowance does not need to be an exact width. Start and finish your tacking on the right side of the fabric (for easy removal of the tacking stitches later). Do not turn under the superfluous fabric at acute angle corners. These 'ears' can simply be avoided, and left on the wrong side of the work, when stitching the pieces together.

3. Assemble your tacked pieces, according to the pattern and their labelling.

4. Sew the pieces together – one possible order is suggested – by placing the pieces right sides together and whip stitching the edges together, taking up a very tiny piece of fabric for each stitch. Use a matching thread (or a co-ordinating fine silk thread) for the least noticeable stitches.

5. Press carefully, especially when pressing the outside edges under.

   Trim, and fold under, any fabric which is visible beyond the outside line. Remove the papers. Tack down the folded under outside edges. Set aside.

6. Cut a 6½″ square of Fabric 2.

7. Cut 1 strip 2″ x 11″ of Fabric 3A (light)

   Cut 2 strips 2″ x 11″ of Fabric 3B (medium)

   Cut 1 strip 2″ x 11″ of Fabric 3C (dark)

   11″ is a little over-long, but leaves a safe margin for completing the mitred corners.

8. Sew the light strip to the lower edge of the central background piece – sewing from and to a point a quarter of an inch from the outer edge of the background piece.

   Sew the dark strip to the top edge, in the same way.

   Sew the medium strips to the sides, in the same way.

   Press from the background towards the frame.

9. Join these corners with mitred seams. It is important here, for the illusion of depth, that the corners are mitred. See picture of quilt example. Press these seams open. This inner square should measure 9½″ (including seam allowance).

10. To make the first narrow border (from Fabric 4):

    Cut two strips 1″ x 9½″. Sew these to the sides of the central framed piece. Press.

    Cut two further strips 1″ x 10½″. Sew these to the top and bottom of the central framed piece. Press.

    Making these corners of the square type makes a break from the series of mitred corners, and gives the impression of a tiny 'flat' area around the framed piece.

    This square should measure 10½″ (including seam allowance). Mark the exact points where the seam lines intersect at each corner.

11. The Icosahedron can now be appliquéd to its background. Experiment with placement, until you find one you think gives the best three-dimensional effect. I machine appliquéd mine with invisible thread.

## The Side Panels – Four the same – *Fig.14,ii (PP)*

**12.** Trace the eight Flying Geese pieces onto paper suitable for Foundation Paper Piecing. Make these up. Press. Trim around them leaving a quarter inch seam allowance. Mark the corner points where the seam lines intersect. Remove the paper. (Scoring every seam with the back of your stitch ripper, along a ruler, makes paper removal very easy). Set aside.

**13.** Make templates for and cut out the four plain pieces used in each side panel, adding a quarter inch seam allowance. Mark the corners where the seam lines intersect. These pieces should be labelled, 'a–d' for example, to identify their positions in the panel. Put the four sets aside.

**14.** Trace four Stars onto freezer paper. Number each piece and label it with the fabric which will be used for it.

**15.** Cut out this Star accurately, and cut the component pieces apart. Press each piece to the wrong side of its fabric, so that it is firmly stuck (albeit temporarily). Sew together – *Fig.14,iii*, matching corner points and pinning only within the seam allowance. Use the edge of the freezer paper as your guide to sewing line. Press each seam as you go. Press from background to Star whenever possible.

**16.** Join the plain panels to the Flying Geese and the Star panels – to the right of the right-hand-side Flying Geese panel, and the left of the left-hand-side panel; and to both sides of the Star panel. Join these three sections together to make each side panel. Trim off any superfluous 'ears' of fabric.

**17.** Noting your marked corners, and matching them, sew the panels to the four sides, from corner to corner. If there is any slight discrepancy

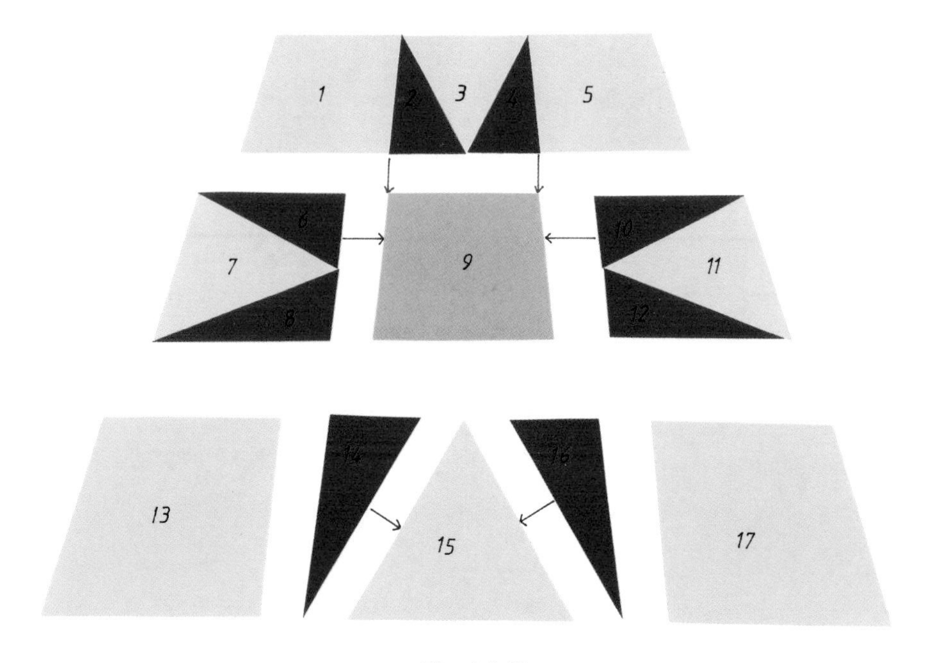


*Fig.14,iii*

in size, preventing the corners from lying in exactly the right position for an accurate mitre, it is possible to make slight adjustments at this stage, because the pieces being joined are plain panels. Once corrected, if necessary, sew the mitred corner seams.

- If you plan to use this centre piece as a block in a large quilt, such as those described in Section 3, your work is complete.

- If, however, you are making a wallhanging or similar small quilt, Finishing Instructions – which are common to all designs – are given at the end of this section.

- Possible quilting designs are given below – *Fig.14,iv.*

- See also full quilting details at the end of this Section.

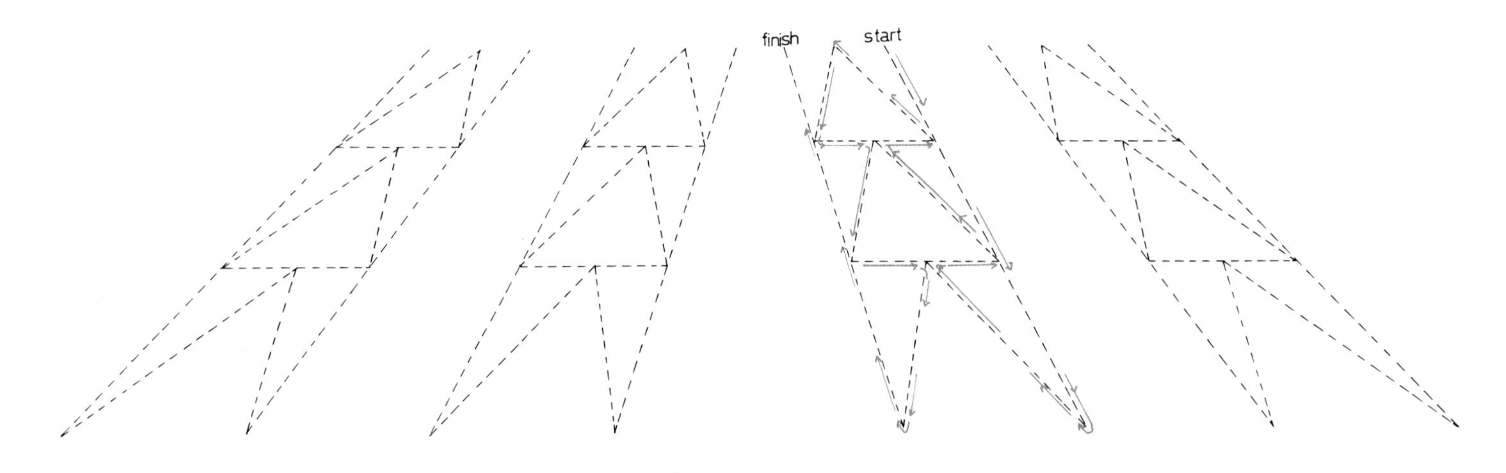


*Fig.14,iv*

# Finishing Instructions

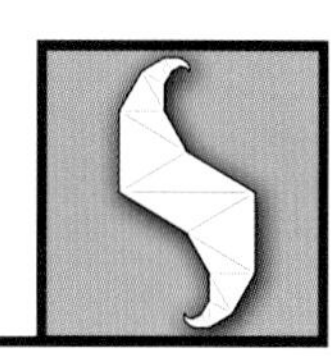

## Applicable to all Quilt Centre Designs

To make your quilt centre into a wallhanging or small quilt in its own right:

First square and trim the centre. Then proceed as follows:

1. Cut strips for the narrow borders, 1" wide x measured height. Sew these to the sides. Press. Measure the width, including the narrow borders, cut two 1" strips at this length and sew to top and bottom. Press.
2. Make four cornerstones using the relevant pattern from the Cornerstone Pattern group in the Pattern Pack, or one of your own choosing. Press.
3. Measure the height and width of the piece including the narrow borders. Cut strips for the wider borders: 2½" wide x measured height and width.
4. Add a cornerstone to each end of the top and bottom strips. Press.
5. Attach the wider borders to the sides. Press.
6. Attach the top and bottom strips. Match the seams exactly, so that the cornerstones fit precisely above and below the side borders. Press.
7. Add batting/wadding and backing to the Top (allowing an inch or so extra all the way round). Baste in your preferred way, and quilt. A quilting suggestion is given for each design.
8. Trim and square the piece; bind with a quarter inch binding, using your own preferred method. I use a single, continuous straight strip – mitred. This needs to be cut at 1¼".

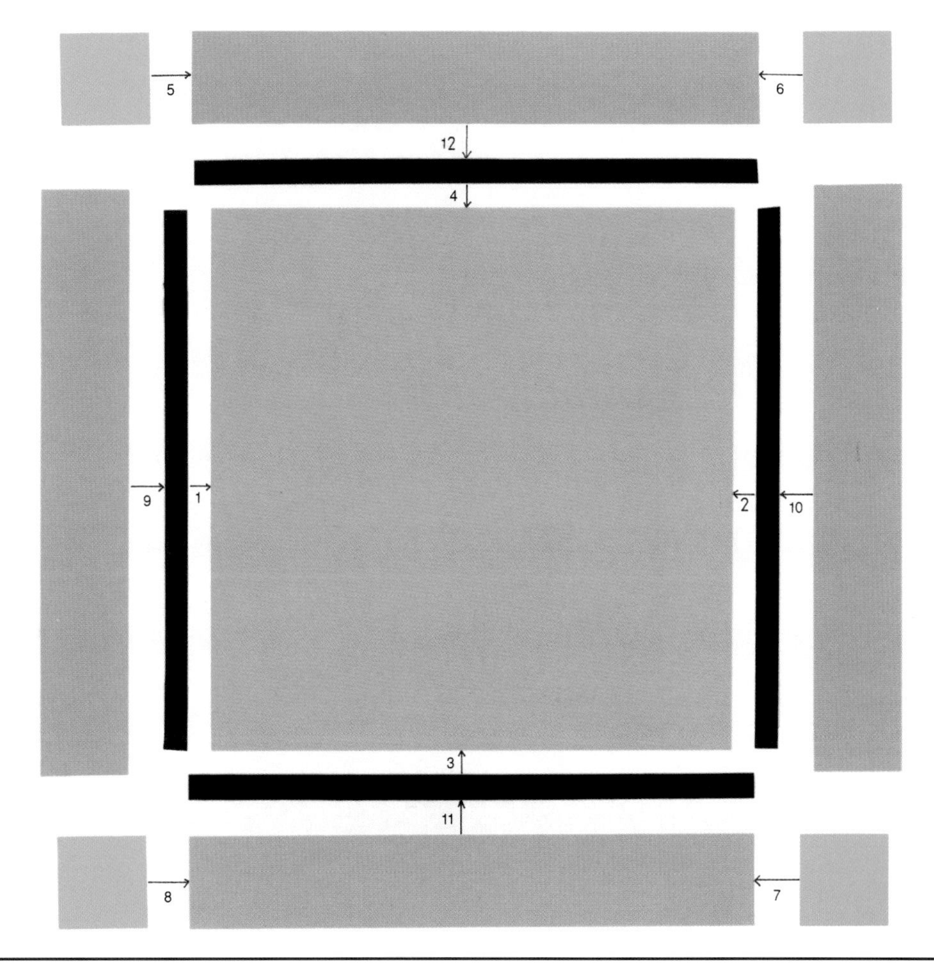

# Further Quilting Notes for Each Small Quilt

## Spiral Triangles

**Narrow Border** Stitch in the Ditch, along both sides, all the way round.

**Cornerstones** Free-motion curved spiral.

**Centre** Quilt along all the curves formed by the twisted log cabin pattern.

Stitch in the Ditch around the hexagon.

Corners – triangular, straight line spiral, as pattern. The distance between the lines can be fixed by using the presser foot as a guide.

**Wider Border** Free-motion irregular feathers, as pattern.

You may mark the 'spine', as shown, if you wish, or do the whole thing freehand.

## Logarithmic Spiral

**Narrow Border** Stitch in the Ditch, on both sides, all the way round.

**Cornerstones** Stitch in the Ditch

**Centre** Free motion spirals (perhaps 8 or 10), from the centre, roughly following the direction of the crossways seams in the wedges.

Corners – free motion daisy-like flower, with trailing leaf stems on either side.

**Wider Border** Diamonds, as pattern. The repeat distance given is 2".

As the centre spiral pattern is appliquéd onto the background, accurate final trimming of the central part will mean that the border pattern can be made to fit exactly. My example was cut at 17½" x 19½".

## Baravelle Spiral

**Narrow Border** Stitch in the Ditch, along both sides, all the way round.

**Cornerstones** A simple ellipse shape, as pattern, placed at right angles to the bias folded ellipse.

**Centre** Free motion spirals, approximately following the outer edges of all the six spirals.

Corners – triangular, straight line spirals, as pattern. The distance between the lines can be fixed by using the presser foot as a guide.

**Wider Border** Free-motion 'stones/dry stone wall' effect, as pattern. The quilting line can be made continuous in each border, by starting at one end of the border and moving down to the next row of 'stones' at the opposite end. I have shown 3 rows of chunky stones. You might prefer 4 rows of flatter ones.

## Archimedean Spiral

**Narrow Border** Stitch in the Ditch, along both sides, all the way round.

**Cornerstones** Square 'spiral' as pattern.

**Centre** Stitch in the Ditch along all the seams of the spiral.

Concentric arcs, as pattern, using the presser foot to maintain a regular distance between. These lines can be made continuous for each half of each corner, by starting where shown by arrow, and following the 'route' suggested.

**Wider border** Free motion zig-zag-type lines, as pattern. These can be made continuous, for each side, by starting at one end and coming back in the opposite direction from the far end.

## Sierpinski Triangle (or Gasket)

**Outline Strip** Stitch in the Ditch along both sides.

**Narrow Border** Stitch in the Ditch, on both sides, all the way round.

**Cornerstones** Stitch in the Ditch

**Centre** Main Triangle – Stitch in the Ditch around the large and medium-sized triangles.

Quilt the motifs, as pattern, on the large and medium-sized, unpieced triangles.

Background – Quilt a triangular grid, based on the small-sized triangle. This will be in the ditch for some, but will go across the surface of the large and medium-sized triangles.

**Wider Border** Fans of leaf/petal motifs, as pattern. This can be a continuous line for a whole border.

The repeat distance given is 3¼", but this can easily be adjusted for the length of your borders, by extending or shortening the curved line between the motifs. My example has 6 on the longer side and 5 on the shorter.

## Pythagorean Tree

**Centre Background** Quilted before the appliqué of the 'tree': hanging diamonds pattern, with twin needles and two different threads, one set of horizontal lines, one of diagonal lines from top left to bottom right.

**Narrow Border** Stitch in the Ditch, on both sides, all the way round.

**Cornerstones** Stitch in the Ditch.

**Centre** Machine appliqué of the 'tree' to the quilted background – straight stitching on all outside edges – also acts as additional quilting.

**Wider Border** Repeated (7), stylised trees (similar to cornerstones), as pattern; continuous line along a whole side.

## Von Koch's Snowflake

**Narrow Border** Stitch in the Ditch, on both sides, all the way round.

**Cornerstones** Need no quilting.

**Centre Design** Stitch in the Ditch to outline the Snowflake.

For each 'petal', free-motion 'fine, waving leaves', as pattern - 2 or 3 'blades' as you prefer. Centre six can be done in a continuous line.

On the triangle divided into trapezia, curvy design, as pattern. This can be done as a continuous line, as the mid-points of the lines are joined.

Four snowflakes, as pattern, in the corners.

Background may be filled with McTavishing filler pattern; or you could use any filler stitch.

**Wider Border** Gentle meandering stitch (not a tight or heavy stippling).

## Lutes of Pythagoras

**Narrow Border** Stitch in the Ditch, on both sides, all the way round.

**Cornerstones** Stitch in the Ditch

**Centre** Stitch in the Ditch around every Lute, every pentagon and every five-pointed star.

Background – Free motion loops and five-pointed stars, as pattern. Follow the inset guide in quilting a five-pointed star freehand. Hint: practise first with pencil and paper until you can make the stars almost without thinking.

**Wider Border** Flattened hexagons, as pattern. The repeat distance given is 1¾". However, it is also necessary to reserve ½" of the length of each border to accommodate the final parts of the pattern. My example has 10 repeats along each side. As the Lutes are appliquéd to the background, accurate final trimming will mean that the border pattern can be made to fit exactly. This centre part needs to be 19" square.

## Rotational Symmetry

**Narrow border** Stitch in the Ditch, along both sides, all the way round.

**Cornerstones** Stitch in the Ditch.

**Centre** 'Leaf' motif, as pattern, in each wedge. The motif overlaps into the colour-matching piece of the adjoining wedge.

Corners – free motion 'loops', as shown in pattern. This can be made a continuous line for each corner by following the suggested 'route'.

**Wider Border** Zig-zag design, as pattern. The repeat distance given is 2". You may need to adjust this to fit your own quilt (See "Before you Begin"). Nine repeats fit on each side of my example.

## Reflective Symmetry

**Narrow Border** Stitch in the Ditch, along both sides, all the way round.

**Cornerstones** Stitch in the Ditch

**Centre** Stitch in the Ditch around the dodecagon.

Corners – Straight lines, as pattern.

Wedges – Stylised flowers, as pattern. One image is given, but you will also need its mirror image, then alternate these designs.

**Wider Border** Free motion flowers, with trailing stems.

## Penrose Tiling

**Narrow Border** Stitch in the Ditch, on both sides, all the way round.

**Cornerstones** Stitch in the Ditch

**Centre** Tiling – Stitch in the Ditch around all of the major pattern lines. If, in your opinion, the hand stitching joining the small pieces is too clearly seen, I suggest using a very, very narrow zig-zag stitch. Using a variegated thread adds a real sparkle to this.

Background – meandering stems and leaves, as pattern.

**Wider Border** Interlocking diamonds, as pattern. The repeat distance given is 2". In my example, there are 10 repeats along each side. As the centre pattern is appliquéd onto the background, accurate final trimming of the piece will mean that the border pattern will fit exactly. My example was cut at 19½" square.

## Tessellation

**Narrow Border** Stitch in the Ditch, on both sides, all the way round.

**Cornerstones** Stitch in the Ditch.

**Centre** First – Le Moyne Star, as pattern, centred precisely on the centre of the pieced blocks, with the outermost points of alternate diamonds placed exactly on the horizontal and vertical seams of the piecing. The length of the diamonds is 3" – equal to the size of the 'H' blocks.

Second – pairs of long Flying Geese, as pattern, are placed at the 'N', 'S', 'E' and 'W' of the Star, and extend to the narrow border.

**Wider Border** Flying Geese, as pattern, with a diamond at the centre of each side, so that the geese are symmetrical. In my example, there are 10 geese on either side of the central diamond, on all four sides. If you need to make a small adjustment to the size, it should be made to the central diamond rather than to the geese.

## Fibonacci Series

**Narrow Border** Stitch in the Ditch, along both sides, all the way round.

**Cornerstones** Stitch in the Ditch

**Centre** Stitch in the Ditch along all strip seams.

Concentric quadrants of a circle, as pattern, starting in each corner – first, and then on either side of it, in the same way. Ultimately these arcs will join up to make a curved octagon. I did not continue these right into the centre, as the strip quilting there was quite close, but that would be an option. Use a freezer paper template as a quilting guide for the first, small quadrant; the remaining arcs may be sewn using the presser foot to keep them a regular distance apart.

**Wider Border** Free motion 'pebbles', as pattern. These can be sewn with a continuous line, for each side. Arrows on the pattern show one way of doing this.

## Perspective and Optical Illusion

**Innermost Border** Stitch in the Ditch, on both sides, all the way round.

**Narrow Border** Stitch in the Ditch, on both sides, all the way round.

**Cornerstones** Stitch in the Ditch to highlight the pyramids.

**Centre Piece** Diagonal grid on the background piece.

Stitch in the Ditch along all edges of the Icosahedron.

Twin needle parallel lines on all four framing strips.

**Perspective Sides** Stitch in the Ditch along all joining seams.

Stitch in the Ditch along all seams of the pieced Flying Geese and the Star.

Flying Geese, as pattern, in the four blank pieces of each side.

**Wider Border** Two 'lines' of a serpentine stitch (or parallel wavy lines, if you do not have fancy stitches on your machine, or you are quilting by hand).

# Section 2 – Facts and Figures

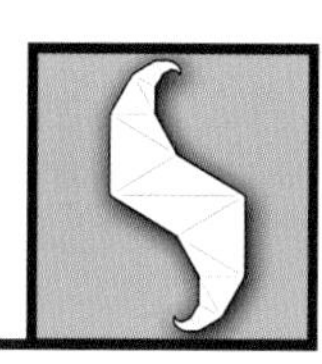

## Guide to Bed/Mattress and Quilt Sizes

| | Mattress Sizes (in inches) | Quilt Size with 16" Drop on three sides and no Pillow Tuck (in inches) |
|---|---|---|
| Single (UK) | 36 x 75 | 68 x 91 |
| Twin (US) | 39 x 75 | 71 x 91 |
| Double (UK) | 54 x 75 | 86 x 91 |
| Full (US) | 54 x 75 | 86 x 91 |
| King (UK) | 60 x 78 | 92 x 94 |
| Queen (US) | 60 x 80 | 92 x 96 |
| King (US) | 78 x 80 | 110 x 96 |

## Guideline sizes for quilts other than Bed Quilts (in inches)

| | |
|---|---|
| Cot/Crib Quilt | from 30 x 42 |
| Lap Quilt | from 54 x 54 |
| Wheelchair Lap Quilt | 36 x 36 |
| Sofa Throw | similar to double / full bed quilt |

## Approximate Conversion from Imperial to Metric measurement system:

| | |
|---|---|
| 1" | 2.54cm |
| 12" / 1ft | 0.3m |
| 3ft / 1yd | 0.9m |
| 4ft | 1.2m |
| 5ft | 1.5m |
| 6ft / 2yds | 1.8m |

# Section 3

Although these small quilts were designed as wallhangings, each to show a geometrical principle, it is possible to use the centres in the same way as blocks, to make bed or other large quilts.

This section shows how, with four completely different styles of quilt; also giving details of quilt sizes that can be achieved by the various methods, with two different sized blocks and sashing widths.

## Size adjustments of individual quilt centres prior to use as blocks in large quilts 1 – 3

As these small quilts were designed primarily to illustrate the geometrical topics, their size and shape were not priority concerns, although I did have an approximate finished size of 24″ in mind throughout (including borders).

Bed/large quilt designs 1 – 3 require that the blocks are square and the same size as one another. The squared size of these small quilt centres is your choice, but I will set out, below, methods of conversion to 18″ and 20″ square blocks. Design 4 illustrates a way in which the blocks may be used in their unaltered state. The size differentials are compensated for by using filler strips of various types and sizes – pieced and unpieced – rather than regular sashing.

### A

Three appliquéd centres require backgrounds which are larger than 18″ square. These, therefore, can only be used to make 20″ blocks.

They are:

- Archimedean Spiral
- Logarithmic Spiral
- Penrose Tiling

To make these into 20″ *(finished)* blocks, appliqué the central patterns to a background piece approximately 21½″ square *(cut)*. Once the appliqué has been completed and pressed, this should then be trimmed to exactly 20½″ square, to give a finished size of 20″ square.

## B

Six of the quilt centres have been designed to finish at 18″ square. They can, therefore, be used, without amendment, as 18″ blocks. They are:

- Rotational Symmetry
- Reflective Symmetry
- Tessellation
- Pythagorean Tree
- Von Koch's Snowflake
- Perspective and Optical Illusion

If you need these for 20″ blocks, add 1½″ *(cut)* strips to all four sides.

## C

The Lutes of Pythagoras design is appliquéd to a background which can be the size of your choosing. Having decided upon the size of block you want, cut your background piece at 19½″ or 21½″ square – depending on whether you want an 18″ or 20″ block *(finished size)*. Trim to an exact 18½″ or 20½″ (these measurements include seam allowances) after the central piece has been appliquéd and pressed.

## D

The Fibonacci Series design is fixed to finish at 17″ square.

This simple pattern represents, by the widths of the strips, from the centre, the famous Fibonacci Series: 1, 1, 2, 3, 5, 8, 13, 21, 34 ...

I have used a unit of half an inch.

If you plan to use this quilt centre as a block on its own, repeated or in a medallion quilt of your own design, no size adjustment is necessary. The medallion quilt illustrated later in this section is based on a block of 18″ *(finished)*.

If a size increase is needed, simply adding strips – as described elsewhere in this section – is not an option, because of the significance of the strip widths in the pattern itself.

Narrow, pieced extension strips, on the other hand, will clearly not be part of the series. The strips will need to be 1″ *(cut)* to make an 18″ *(finished)* block; and 2″ *(cut)* for a 20″ *(finished)* block: added to all sides. The diagram below shows a simple example of how this can be done.

*Example of pieced extension strip*

# E

The remaining quilt centre designs are pieced into their backgrounds and will need to have strips added to them to achieve an 18″ or 20″ square *finished size*. They are:

- Spiral Triangles
- Baravelle Spiral
- Sierpinski Triangle

**i) Spiral Triangles** – 16″ wide x 18½″ high (including seam allowances)

To achieve an 18″ *(finished)* square/block:

Nothing needs to be added to the height.

Cut 2 strips of background fabric 1¾″ wide x 18½″ long.

Sew these to the two sides. Press.

To achieve a 20″ *(finished)* square/block:

Cut 2 strips of background fabric 2¾″ wide x 18½″ long.

Sew these to the sides. Press.

Cut 2 strips of background fabric 1½″ wide x 20½″ long.

Sew these to the top and bottom. Press.

**ii) Baravelle Spiral** – 18½″ wide x 16″ high (including seam allowances)

To achieve an 18″ *(finished)* square/block:

Nothing needs to be added to the sides.

Cut 2 strips of background fabric 1¾″ wide x 18½″ long.

Sew these to the top and bottom. Press.

To achieve a 20″ *(finished)* square/block:

Cut 2 strips of background fabric 1½″ wide x 16″ long.

Sew these to the sides. Press.

Cut 2 strips of background fabric 2¾″ wide x 20½″ long.

Sew these to the top and bottom. Press.

**iii) Sierpinski Triangle**

The basic triangle is 18″ wide and 15½″ high *(finished)*.

If the pieced background triangles are added to this, without the half-inch framing strip on the sides of the basic triangle, the piece will be the same size as the Baravelle Triangle, and the size adjustments will be the same.

Therefore, if you are planning to use this design as a block in a large quilt, I suggest you omit the half-inch strip.

If you do wish to leave it in, then measure your own centre when it is finished and add strips as necessary to achieve the size you want to work with.

# Design 1 – STRAIGHT SET – 4 x 4

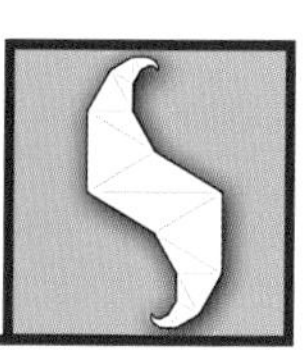

*The example shown is with 20″ blocks and 3″ sashing, finished sizes, for use with UK King Size/US Queen Size beds. 16 blocks are needed.*

**Use:**

**i.** Same block repeated;

**ii.** Same block, made in two colourways, alternated;

**iii.** Two different blocks alternated;

**iv.** Selection of different blocks;

**v.** Borders, to customise size.

## Assembly Instructions

1. Square all blocks as necessary and arrange as desired.
2. Add short sashing pieces to the right hand side of the first three blocks in each row.
3. Join these three blocks plus the fourth one to form the rows.
4. Add a long sashing strip to the lower edge of the top three rows.
5. Add the four outside long sashing strips. The corners may be square or mitred.

*Assembly Diagram*

i. If square: add side sashings, then tops – which must be cut long enough to include the width of the side sashing strips.

ii. If mitred: allow an extra 5" at both ends of each strip. Add the four sashing strips in sequence **finishing the seams ¼" before the ends.**

6. Add borders, if necessary, to make the quilt exactly the size you wish it to be.

**Guide to yardage for straight set sashing strips**

Based on: 20" block, 3" sashing, 41" usable width of fabric

**2.5m (2¾ yds)**

To allow for: mitred outer corners, seam allowances, variations in cutting, usable width less than 41".

Cut lengthways. Outer frame strips will have a join.

## Quilt Sizes Achieved With A Straight Setting

| 18" Blocks | | |
|---|---|---|
| Arrangement | 2" Sashing | 3" Sashing |
| 2 x 2 | 42 x 42 | 45 x 45 |
| 2 x 3 | 42 x 62 | 45 x 66 |
| 3 x 3 | 62 x 62 | 66 x 66 |
| 3 x 4 | 62 x 82 | 66 x 87 |
| 4 x 4 | 82 x 82 | 87 x 87 |
| 4 x 5 | 82 x 102 | 87 x 108 |
| 5 x 5 | 102 x 102 | 108 x 108 |

| 20" Blocks | | |
|---|---|---|
| Arrangement | 2" Sashing | 3" Sashing |
| 2 x 2 | 46 x 46 | 49 x 49 |
| 2 x 3 | 46 x 68 | 49 x 72 |
| 3 x 3 | 68 x 68 | 72 x 72 |
| 3 x 4 | 68 x 90 | 72 x 95 |
| 4 x 4 | 90 x 90 | 95 x 95 |
| 4 x 5<br>(5 x 4) | 90 x 112<br>(112 x 90) | 95 x 118<br>(118 x 95) |
| 5 x 5 | 112 x 112 | 118 x 118 |

The above sizes may be customised to match your own needs by the addition of border. Remember that, for a bed quilt, borders may be on sides and lower edge only; or there may by different widths on sides/top/bottom.

Comparison with the figures given in the Table in Section 2 will show the arrangement which will result in a quilt most closely suited to the size you want.

# Design 2 – DIAGONAL SET/ON POINT – 3 x 3

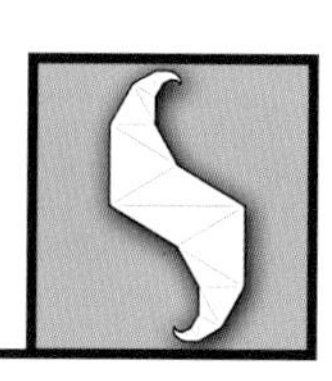

*Illustration shows on point setting with 20" blocks and 3" sashing (finished sizes). 13 full blocks are needed, 8 side setting triangles, 4 corner setting triangles.*

**Use:**

**i.** Same block repeated;

**ii.** Same block, in two colourways, alternated;

**iii.** Two blocks, alternated;

**iv.** Different blocks.

Some of the Geometrical Quilt centre blocks are not well suited to a diagonal setting, because they only have one axis of symmetry. They include:

Sierpinski Triangle and Pythagorean Tree.

Lutes of Pythagoras is completely asymmetrical, so it can be set straight or diagonally.

## Assembly Instructions

**1.** Square individual blocks if necessary and arrange.

**2.** Add short sashing pieces to both sides of blocks 1, 2, 5, 10 and 13.

**3.** Add short sashing pieces to the right hand side of blocks

3, 4, 6, 7, 8, 9, 11 and 12.

**4.** Join blocks 2, 3 and 4 - row 2

5, 6, 7, 8 and 9 – row 3

10, 11 and 12 – row 4

**5.** Add side setting triangles, as follows:

A and H to Block 1/row 1

B and G to row 2

C and F to row 4

D and E to Block 13/row 5

**6.** Add first corner setting triangles, as follows:

x and z to row 3

7. Add longways sashing strips, cut to match row length, to the top edges of rows 1, 2, and 3; and also to the lower edges of rows 3, 4 and 5.

8. Add further corner setting triangles, as follows:

   w and y to row 1's and row 5's sashing strip - centred carefully on Blocks 1 and 13.

9. Join the rows together, centering carefully so that Blocks 1, 3, 7, 11 and 13 + corner triangles w and y form a straight row beneath one another.

10. Add outer sashing strips/narrow borders. With a diagonal setting, mitred corners will give a better finished appearance. So, add between 6" and 8" to the measured length of each strip to allow for the making of neat mitres; and start and stop sewing ¼" before both ends of the piece to which they are being sewn.

11. Add borders, if required, to customise the quilt size for your own purposes.

## Approximate Quilt Sizes Achieved

### Diagonal Setting with 18" or 20" Blocks and 2" or 3" Sashing. 2" or 3" border/sashing on all sides

There are not as many useful sizes which can be made with this configuration. However, as it looks so attractive, it is well worth considering. The illustration shows a 3 x 3 setting.

| 18" Blocks | | |
|---|---|---|
| **Arrangement** | **2" Sashing** | **3" Sashing** |
| 2 x 2 | 64 x 64 | 70 x 70 |
| 2 x 3 | 64 x 92 | 70 x 100 |
| 3 x 3 | 92 x 92 | 100 x 100 |

| 20" Blocks | | |
|---|---|---|
| **Arrangement** | **2" Sashing** | **3" Sashing** |
| 2 x 2 | 70 x 70 | 76 x 76 |
| 2 x 3 | 70 x 101 | 76 x 108 |
| 3 x 3 | 101 x 101 | 108 x 108 |

### Guide to Yardage for Sashing on Diagonal Set Example

Based on 20" Block, 3" sashing, 41" usable fabric width: you need 3.25 m (3½ yds).

To allow for mitred corners on outside corners, seam allowances, variations in cutting, usable width less than 41". Cut strips lengthways for sashing with no joins.

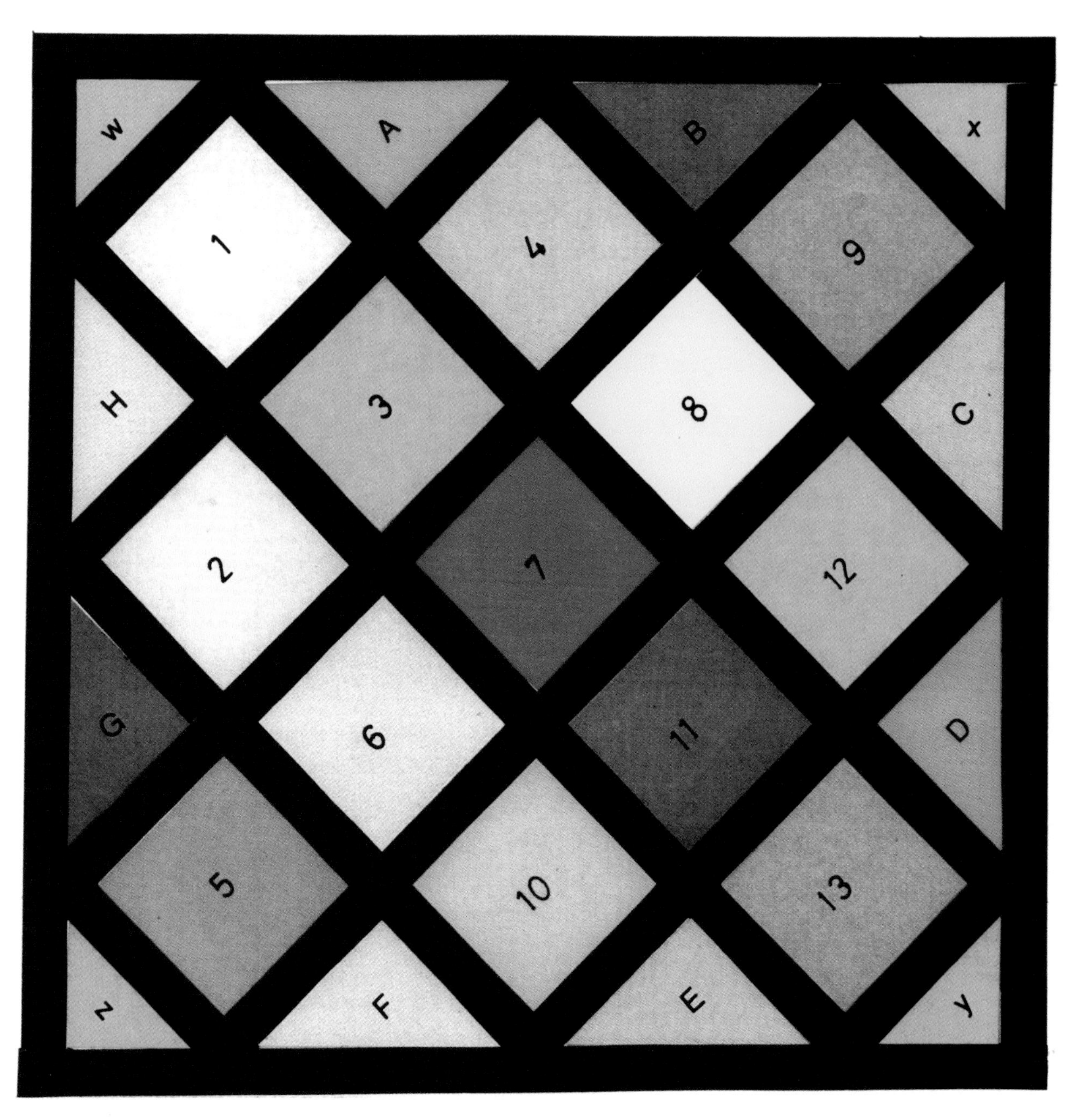


*Assembly Diagram*

Cutting lengthways, at 3½" wide, you will need:

4 cuts for the outer frames

2 cuts for the long sashing strips

3 cuts for the short sashing strips

2 cuts for the medium and two horizontal short sashing strips

**Guide to Yardage for Side and Corner Setting Triangles for Diagonal Set Example**

*(The square dimensions given are a little on the generous side. Final trimming will be needed).*

In order to have the straight grain of the fabric on the outer edges, side triangles need to be cut as quarter squares; corner triangles need to be cut as half squares.

**For 20" Blocks:**

The side triangles are cut from two 30" squares, cutting on both diagonals.

The corner triangles are cut from two 15¼" squares, cutting on one diagonal only.

You need 2m (2¼ yds)

Cut the two smaller squares side by side across the width of the fabric. Cut the larger squares side by side along the remaining length of the fabric.

# Design 3 – MEDALLION – based on a 18" centre block

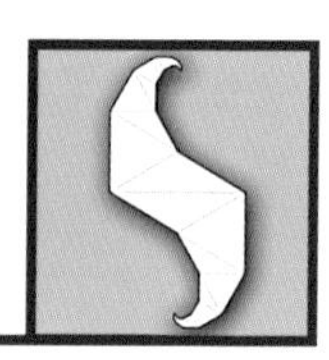

*The blocks best suited for a Medallion Quilt, are:*

Rotational Symmetry

Reflective Symmetry

Penrose Tiling

Tessellation

Fibonacci Series

Von Koch's Snowflake

Perspective and Optical Illusion

Medallion Quilts, typically, use a fairly limited colour palette, often based on one of the background fabrics. Traditionally, fabrics and elements of the design are repeated throughout the quilt.

Some patterns, which would be suitable for pieced borders, are given later, as possible filler strips for Design 4. But, numerous block patterns would be suitable for this use. Sizes will have to be adapted to suit your own measurements. In this context, simple patterns are preferable to overly complex ones. A Medallion quilt depends very much, for its success, on overall – rather than individual section – impact.

**Required Elements – for the design illustrated**

**i.** Centre Block.

**ii.** Two further blocks for the pieced corners. These may be of the same design as the centre block, or an alternative design that interacts well with the centre block. However, the corner blocks/half-blocks should be the same design as one another.

**iii.** Two designs, of your choice, for the pieced borders.

**iv.** Two background fabrics for the non-pieced borders and corners surrounding the centre block.

**v.** Contrasting strip for inner border and outer frame. This fabric should contrast with Background 1 and should be used elsewhere in the quilt e.g. somewhere within the pieced elements.

## Assembly Instructions

**1.** Square the Centre Block if necessary. Finished size should be 18″ (+ ½″)

**2.** Cut 4 half-square triangles from Background 1 fabric.

To keep the straight grain of the fabric on the outside edge, these four triangles should be cut from two squares of 13½″. Sew one triangle to each side of the central block. Square again, if need be, after attaching the triangles. It is very important that this quilt should be square at every stage.

**3.** Add a narrow contrasting straight strip all around the square resulting from stage 2. The fabric should be one which has been, or will be, used in the quilt construction, contrasting to Background fabric 1, as far as possible. Sew the strips to the square and mitre the corners.

This strip is functional. It is required to increase the centre block + triangles to a square of 30″ (+ ½″). Mine finished at 25½" (+ ½″), so my strips were 2¼″ (+ ½″) wide. Yours may be slightly different.

**4.** The length of 30″ provides an accurate base for a pieced border of 6″ blocks. Make 24, 6″ blocks in your own choice of pattern. If the patterns bears some resemblance to, or has some visual connection with, the centre block, that will enhance the overall appearance of the quilt. Join the blocks in 2 sets of 5, and 2 sets of 7. Join the sets of 5 to the sides, and the sets of 7 to the top and bottom of the 30″ square.

(If you are using a border pattern, rather than blocks, work to length only)

The bordered square should finish at 42″ (+ ½″).

**5.** Cut four strips from Background fabric 2. These should be cut 9 7/8″ wide and 44″ long.

Sew one to each side of the 42″ (+ ½″) square, centering them carefully. The superfluous corners will be cut off at Step 7.

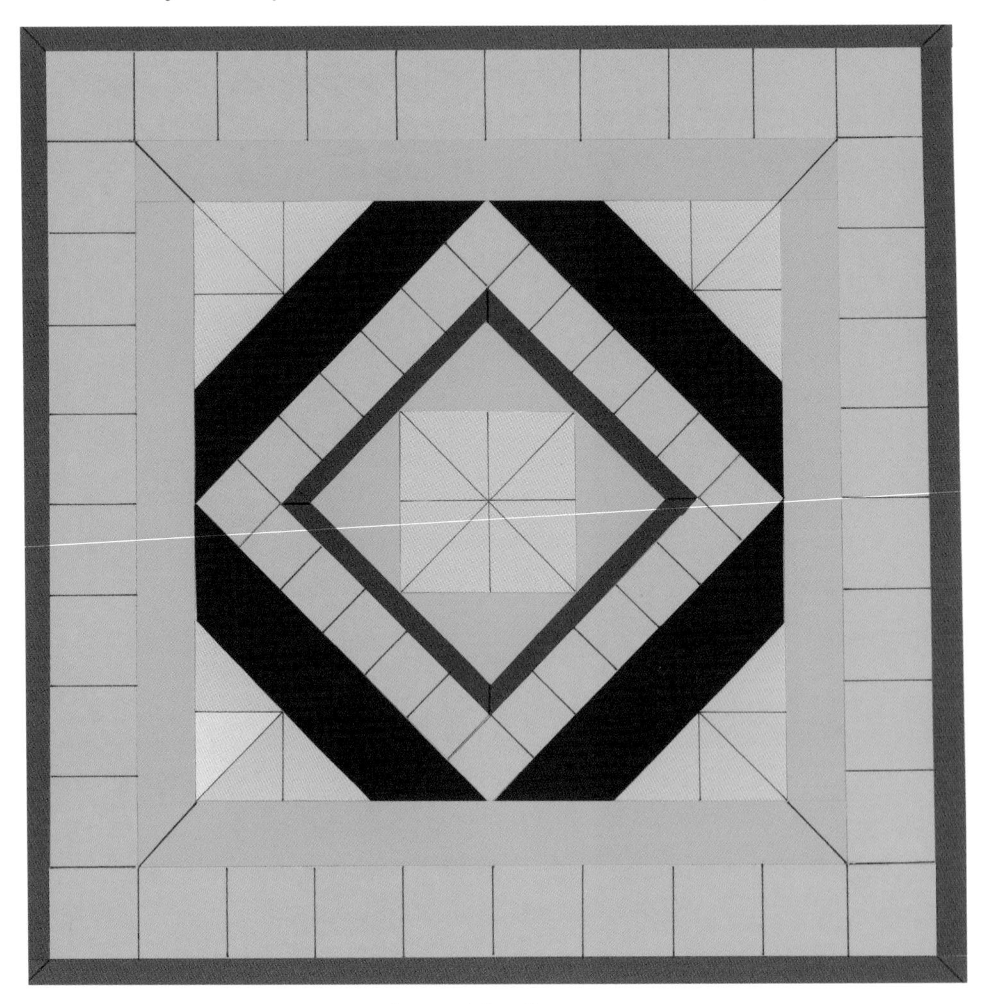

*Assembly diagram*

6. Prepare the pieced corners. Two 18″ (+ ½″) blocks are needed. These blocks may be the same as the centre block, with, perhaps, some colour variation. If you use different blocks, choose one with a similar appearance e.g. the two Symmetry blocks.

   Cutting these blocks along their diagonal and re-attaching them, with a quarter inch seam, will reduce their size slightly. This reduction is compensated for by a small extra width of the previous unpieced border.

   If your chosen block is not four-way symmetrical, e.g. Tessellation, cut one block along the 'top left to bottom right' diagonal; and the other along the 'bottom left to top right' diagonal. Place them symmetrically.

7. Centre these corners carefully and sew one to each Background Fabric 2 strip.

   Re-square the quilt at this stage, cutting off the superfluous corner fabric from the Background 2 strips. It should be 60" (+ ½″) square.

8. Add a 6″ (+ ½″) border, of Background fabric 1, to all sides of this square.

   The resulting square will now be 72″ (+ ½″) square.

9. 72″ is a very convenient length, as it can accommodate blocks of 6″, 9″, 12″ and 18″. Thus, this final pieced border can be used to bring the quilt very close to its desired ultimate size.

   I have shown a border of 9″ blocks.

   Select a pattern which fits well with the patterns so far used.

   Make 36, 9″ blocks.

   Join in two sets of 8 blocks and two sets of 10 blocks. Join the sets of 8 to the sides, and the sets of 10 to the top and bottom of the 72″ square.

   The quilt will now be 90″ (+ ½″) square.

10. The illustration shows a final unpieced border of 3″ (+ ½″) x approximately 100″ (cut). This border will allow for further fine adjustment of size, if this is necessary.

    I suggest using the same fabric as the narrow inner border for this final border.

**Guide to Yardages for the Border elements of the Medallion Quilt.**

Background Fabric 1: **1.75m**

**(2 yds)**

Cut the two squares side by side across the width of the fabric; then eight strips, widthways, across the remaining length

Background Fabric 2: **1.25m**

**(1½ yds)**

Cut the four wide strips lengthways

Contrasting Strips: **1.5m**

**(1¾ yds)**

Cut eight 3½″ strips. One join for each long piece

Cut three stips of your narrow width and re-cut as necessary to fit. One join for one of the short pieces

# Design 4 – OFFSET

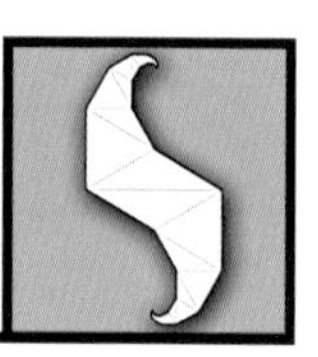

*This would be a fun design to make, and the quilt centres could be used without any size adjustment at all. Everyone's quilt will be unique.*
*As the design is so individual to the maker, it is not possible to give precise instructions. My illustration is an example of one way such a quilt can be made.*
*For this design I show it made to a single/twin size.*

## Assembly Instructions

The construction principles below can be applied to any size of quilt.

1. Decide on the quilt size you want. Draw this to scale on a large sheet of paper. I used a scale of 1:10, as this is a simple scale to work with – whether you are calculating from 'actual' to 'scale' or vice versa. Although graph paper is usually a marvellous aid, I recommend using plain paper for this drafting exercise, as the graph paper lines will distract from the lines in your drawing; and, if you are drawing in pencil (and you will need to correct as you go along), some of the lines will be difficult to see, if you use graph paper. You could use graph paper underneath the plain paper to help with the initial drawing.

2. Select the quilt centres/blocks you wish to use. Measure them carefully, discount seam allowances (i.e. take off ½") and use only the finished sizes in your drawing. These seam allowances must be added back in when you are calculating cutting sizes from the scale drawing. Draw all your blocks to scale on another sheet of paper. Label them with names and sizes. Cut them out.

3. Position them on your scale drawing, until you find a pleasing arrangement. You will see that I used just three blocks for the centre row, between two rows of four. Because the blocks, though different, are of a similar size and shape, I found this the best way to enhance the difference between this arrangement and a regular, straight setting.

4. Stick the cut-out blocks to the scale drawing, in your chosen positions.

5. Where there are large gaps: for example, greater than 3", you can plan to insert a pieced filler strip – shown on the illustration by a hatched rectangle. The narrowest I used was 3", the widest, 8". Draw the possible positions of the pieced strips on your plan. Smaller gaps are filled by sashing-like strips, in a fabric to unify the whole quilt.

6. Work systematically through this first draft, to determine how best to put the various elements together. This stage will probably be the longest part of the design phase.

7. Once it all appears feasible, look it over once again, this time examining it for opportunities to 'tweak' sashing and filler strips – or perhaps move a whole block slightly – so that the whole piece can be assembled into sections. You will see that, in the illustration, it is

possible to put it together in three vertical sections. Sometimes, just dividing a piece of sashing will open up that possibility.

8. This kind of design really does need a 'framing' border – even just a 2" one, as in the illustration – to bring the whole quilt together.

9. Suggestions for filler strip patterns are included later in this section.

*Assembly Diagram – One Possibility*

## Possible Designs for Pieced Borders and Filler Strips for Designs 3 and 4

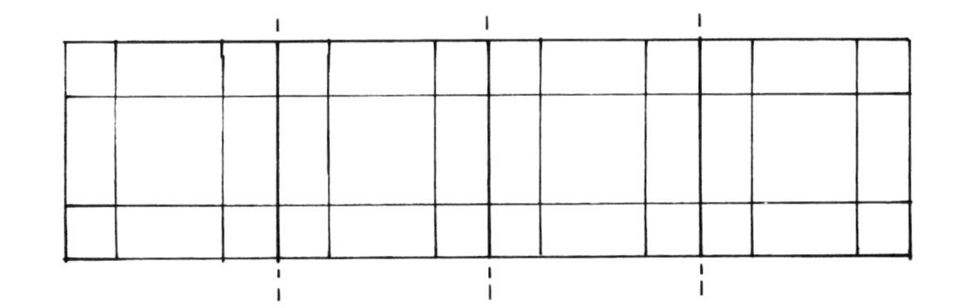

***Uneven Nine Patch***
*May be strip pieced with secondary cuts*

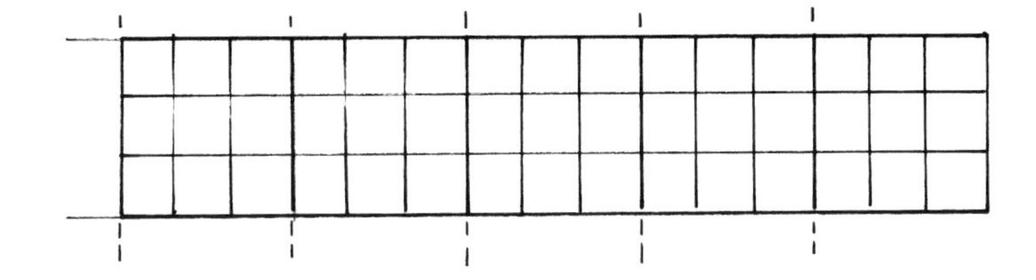

***Nine Patch***
*May be strip pieced with secondary cuts*

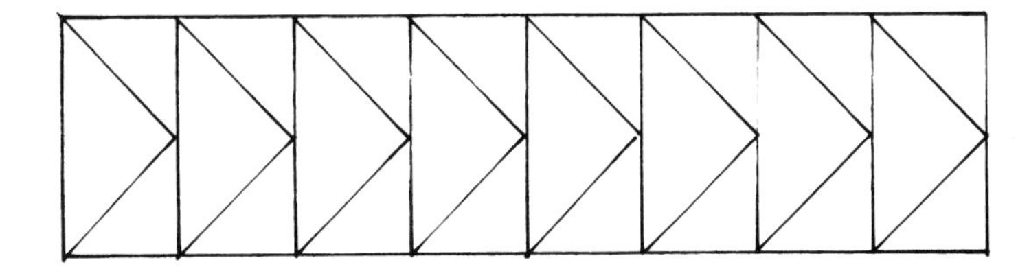

***Flying Geese***
*(common proportions: 'goose' height = half of the strip width*
*May be paper pieced in long strips, or by your own preferred method*

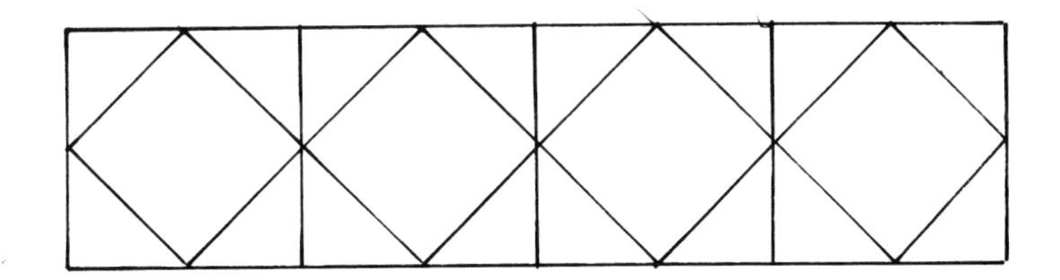

***Square in a Square***
*May be foundation paper pieced individually, or by your own preferred method*

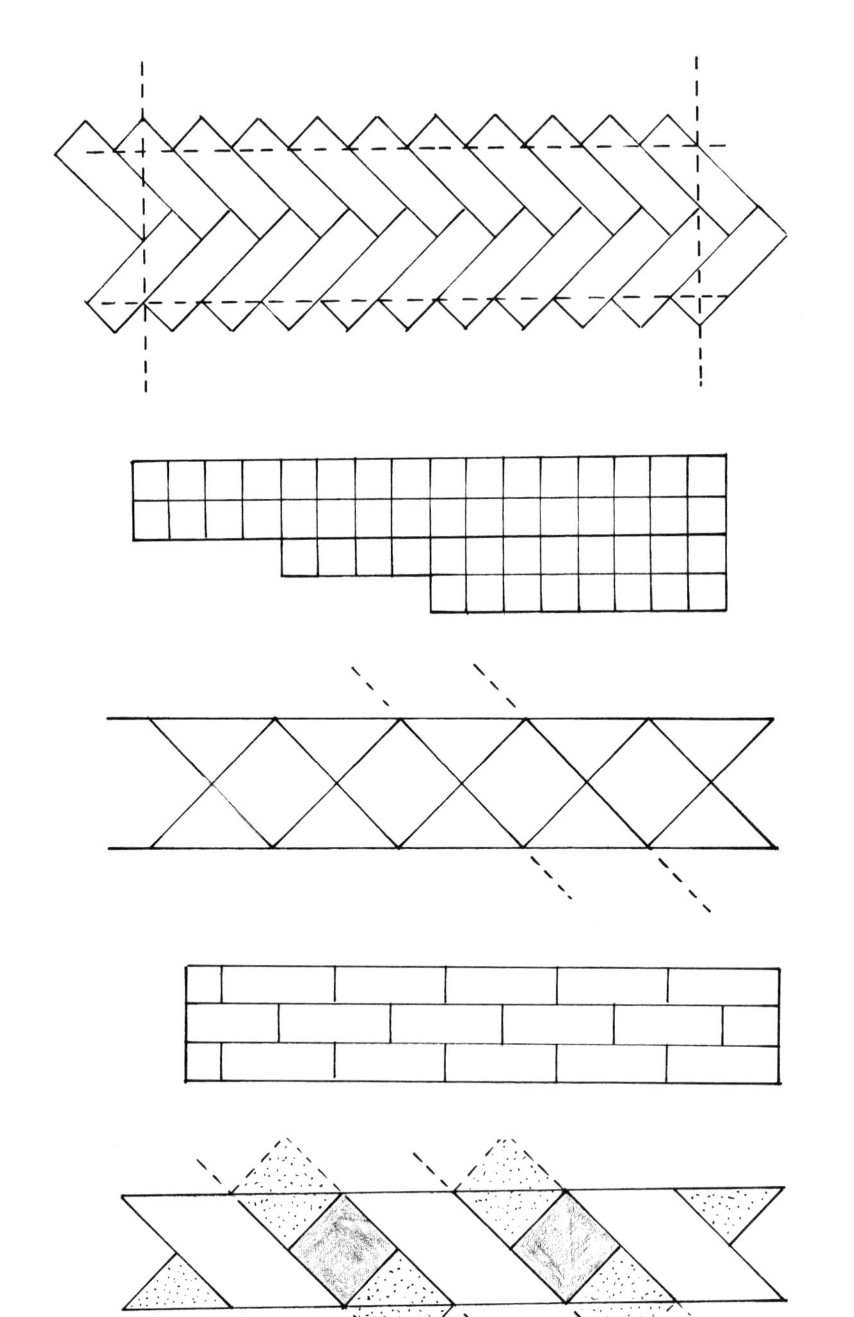

***Herringbone or Braid***
*Short rectangular pieces placed at right angles to one another. May be made by foundation paper piecing, or individually cut pieces*

***Chequerboard***
*May be strip pieced, with secondary cuts*

***Seminole Diamonds***
*Strip pieced, with 90 degree secondary cuts. Set at 45 degrees, then trimmed*

***Brick Pattern – Stretcher Bond***
*May be strip pieced in rows,alternate rows offset by half a brick*

***Ribbon***
*Seminole piecing, with 90 degree cuts, alternating with single strips, set at 45 degrees, then trimmed*

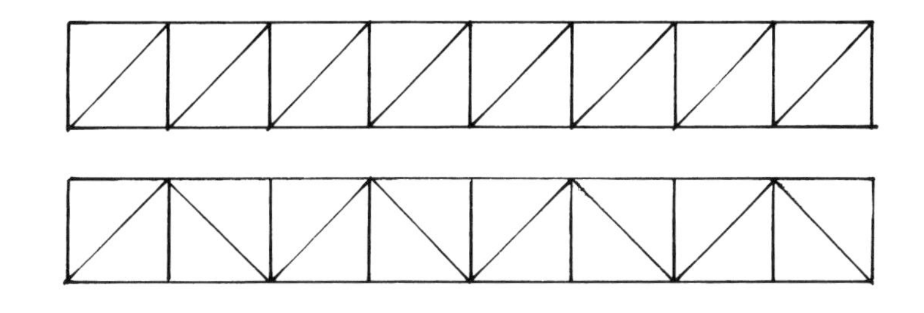

***Half Square Triangles (same direction)***
***Half Square Triangles (alternate direction)***
*May be pieced by pairs of squares, or foundation paper pieced*

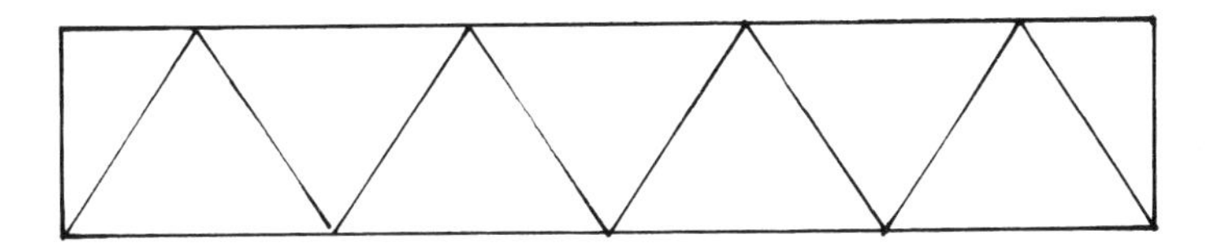

***Isosceles Triangles***
*May be foundation paper pieced in long strips*

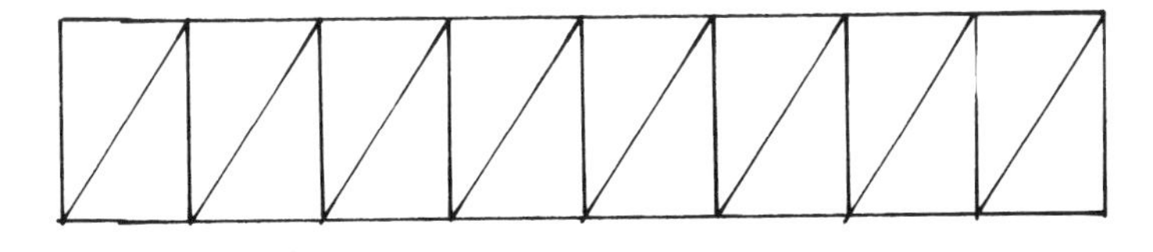

***Sawtooth Triangles***
*May be foundation paper pieced in strips*

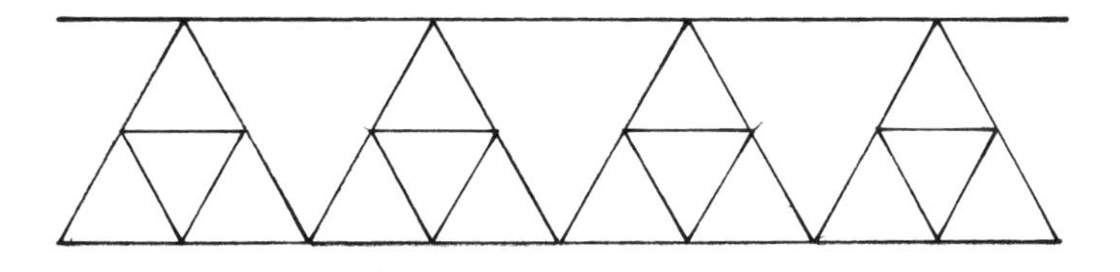

***Equally divided Equilateral Triangles, alternating with Plain triangles***
*Individually foundation paper pieced, alternating with template-cut triangles*

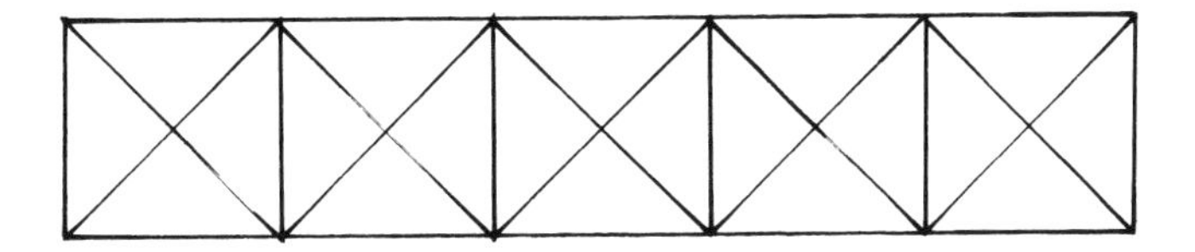

***Quarter Square Triangles***
*Pieced with two sets of pairs of squares (sets of half-square triangles, divided in two, re-paired with different fabric sets and re-joined).*
*3D image can be made by careful colour placement*

# APPENDIX

## Books for beginners

*Your First Quilt Book (or It Should Be!)* Carol Doak
*Quilter's Complete Guide* Marianne Fons and Liz Porter
*Show Me How to Paper Piece* Carol Doak

## A few websites for all

www.quilterscache.com
www.quiltville.com *(Look at menu at left of page for tips, techniques and patterns.)*
www.victorianaquiltdesigns.com *(Many pages of techniques, instructions on topics etc. – not just designs.)*
www.quilting.about.com
www.quilt.com
www.ritasquilts.com *(Look for "Projects and Lessons" and "Tutorials")*
www.vrya.net/quilt/index.php *(For calculating yardage.)*
www.patchpieces.hypermart.net/quiltsettings.html *(For working out setting triangles, etc.)*
www.phancypages.com/newsletter/ZNewsletter2095.htm *(For simple colour theory.)*

## Books I have referred to, directly or indirectly, for this book

*Encyclopaedia of Pieced Quilt Patterns* Barbara Brackman
*Curves in Motion* Judy Dales
*Tucks, Textures and Pleats* Jennie Rayment
*Medallion Quilts* Jinny Beyer
*Pieceful Scenes* Angela Madden
*All-in-One Quilter's Reference Tool* Harriet Hargrave, Sharyn Craig, Alex Anderson, Liz Aneloski
*Mastering the Art of McTavishing* Karen McTavish
*60 Machine Quilting Patterns* Pat Holly and Sue Nickels
*Machine Quilting Made Easy* Maurine Noble

*Plus* numerous Mathematical Websites.